T0276125

Chemotherapeutic Targets in Parasites

Contemporary Strategies

Parasitic infections are the most prevalent of human diseases. Parasites' effective evasion of their hosts' immune defenses and their complex physiology and life cycles make them especially resistant to attack by chemotherapeutic agents. Researchers continue to face the challenge of designing drugs to successfully counteract them.

Chemotherapeutic Targets in Parasites analyzes the critical metabolic reactions and structural features essential for parasite survival and advocates the latest molecular and biochemical strategies with which to identify effective antiparasitic agents. An introduction to the early development of parasite chemotherapy is followed by an overview of biophysical techniques and genomic and proteomic analyses. Several chapters are devoted to specific types of chemotherapeutic agents and their targets in malaria, trypanosomes, leishmania, and amitochondrial protists. Chapters on helminths include metabolic, neuromuscular, microtubular, and tegumental targets. Emphasized throughout is the design of drugs that are more selective and less toxic than those used in the past.

A comprehensive discussion of selective targets in parasites for new drugs is long overdue. This up-to-date book will be especially relevant to medical and clinical researchers and to graduate students in parasitology, pharmacology, medicine, microbiology, and biochemistry.

Tag E. Mansour is Professor Emeritus in the Department of Molecular Pharmacology at the Stanford University School of Medicine. His research on biochemical and molecular parasitology and the action of antiparasitic agents has been published extensively.

Chemotherapeutic Targets in Parasites

Contemporary Strategies

Tag E. Mansour

with the assistance of
Joan MacKinnon Mansour

CAMBRIDGE
UNIVERSITY PRESS

CAMBRIDGE UNIVERSITY PRESS
Cambridge, New York, Melbourne, Madrid, Cape Town, Singapore, São Paulo

Cambridge University Press
The Edinburgh Building, Cambridge CB2 2RU, UK

Published in the United States of America by Cambridge University Press, New York

www.cambridge.org
Information on this title: www.cambridge.org/9780521620659

First published 2002
This digitally printed first paperback version 2005

A catalogue record for this publication is available from the British Library

Library of Congress Cataloguing in Publication data
Mansour, Tag E.
    Chemotherapeutic targets in parasites : contemporary strategies / Tag E. Mansour.
        p.   cm.
    Includes bibliographical references and index.
    ISBN 0-521-62065-1 (hc)
    1. Antiparasitic agents.   2. Parasitic diseases – Chemotherapy.   I. Title.
    RM412 .M365   2002
    616.9′6061 – dc21   2002067057

ISBN-13 978-0-521-62065-9 hardback
ISBN-10 0-521-62065-1 hardback

ISBN-13 978-0-521-01836-4 paperback
ISBN-10 0-521-01836-6 paperback

Contents

Preface

I became intrigued with parasites when I started my research career as a graduate student in England. I was rightly told that the field of parasites has a great future for a starting biochemist/pharmacologist. The field of parasite research was not crowded and therefore was designated as a neglected area of research. Subsequently, a large number of talented and highly sophisticated young scientists were attracted by the urgent need for modern studies on parasites and antiparasitic agents. The fields of parasite biology and biochemistry accumulated a large volume of information that led to the possibility of rational design of antiparasitic agents. There is renewed hope for discovery of more selective and less toxic drugs against parasites.

At the present time infections by parasites, both protozoal and helminthic, constitute the most prevalent diseases in the world. The World Health Organization estimates that there are at least 3 billion people in the world who are infected with parasites. Many of these harbor more than one infection. The prevalence of these infections, especially in developing countries, is not only a cause of untold human suffering and mortality but a growing impediment to better local and global economies.

The prime aim of this book is to discuss critical metabolic reactions and cellular structural features that are essential for survival of parasites, particularly those that differ from those of the host. A comprehensive discussion of selective targets in parasites for old and new drugs is long overdue. The term selective targets may not apply to targets of older antiparasitic agents. Most of these early drugs were not discovered by a rational procedure.

Chapter 1 includes a discussion of the development of parasite chemotherapy from Paul Ehrlich's time to our new era when the search for antiparasitic agents has been influenced by the impact of modern biochemistry and biology. Animal models and *in vitro* cultures for screening are discussed. Traditional ways of designing antiparasitic chemicals to inhibit the functioning of specific targets, the more modern use of relationships between chemical structure and biological activity to design new drugs, and the latest techniques of combinatorial chemistry to prepare hundreds of thousands of new molecules are included.

In Chapter 2 I introduce several recently successful biophysical techniques with which to analyze drug–target interactions. Nuclear magnetic resonance is now used to carry out noninvasive experiments on the energy metabolism of intact parasites. Some aspects of DNA technology are discussed generally, but not in detail, because specialized laboratory manuals are readily available. Mention is made of the latest information about genomic analysis of major parasite groups. Structural genomics is a growing field that promises to have a major impact on identifying three-dimensional structure and function using DNA sequences. Proteomics is another new area that is rapidly expanding. Some scientists whose major interests are in genes seem to have forgotten that the actions of genes are manifested through proteins. This is particularly important for identification of drug targets in parasites.

In Chapters 3–6 I emphasize those aspects of parasite life in the host that have distinguishing features such as metabolic differences between host and parasite. Also included are discussions of the mechanism of action of some of the current antiparasitic agents against their targets. The concept of "selective toxicity" is emphasized and the most selective drugs are more fully described. A few examples of the value of determining the mechanism of action of both old and new drugs are given. It was in the 1950s that the discoveries of Lederberg, and Park and Strominger drew attention to bacterial cell wall synthesis as a selective target for penicillin and other antibiotics. More research on mechanisms of action will lead to discovery of new targets and new antiparasitic agents.

The subject of parasite resistance to certain antiparasitic agents has been integrated with discussion of the mechanism of action. In many cases studies on resistant strains of parasites gave clues to new drug targets with more details about drug–target interactions. There are some divergent views in the literature about parasite resistance. Information on drug resistance from *in vitro* cultures and from laboratory animals should always be considered in relation to human field studies.

In the Chapters 7 and 8 I discuss topics that have been less studied than parasite metabolism. Motility of parasitic helminths plays an important role in maintenance of their location in the host. Many parasites can be eliminated from the host by drugs acting on the neuromuscular receptors of the parasites. Information is given about neuromuscular receptors, neurotransmitters, and changes in parasite behavior as a result of antiparasitic agents. This is an area that has not been fully exploited in the search for new drug targets. Also included are discussions of the microtubules that are basic to control of the location of intracellular organelles and how the benzimidazoles owe their anthelmintic activity to blocking of the microtubular matrix. The last chapter is a discussion of the tegument of platyhelminths as a target. Although there are several anthelmintics that affect components of the tegument, there have not been enough studies of the different ways the tegument functions to help the parasites' life in the host. It is generally accepted that the tegument of flatworms plays an essential role in the transport of nutrients from the host and acts as a protective shield. The interaction between antiparasitic agents against the tegument and the host's immune system has a potential synergistic role in therapy.

Some of the chapters include sections titled "Potential Research." These cover areas that require more experimentation and are included for those who are fortunate enough to be able to go to the laboratory to perform experiments.

The life cycles of several typical parasites are briefly given for the sake of clarity in the discussion of targets. These are not intended as a substitute for more detailed descriptions found in parasitology textbooks.

The bibliographies at the end of each chapter are not comprehensive, but they should give the reader indications of where to look for additional information. References to reviews should be useful to readers who wish to have an overview of a particular area. Although I have been careful to refer to both old and new publications, I may have neglected some of great distinction and so offer my apologies to my biochemistry, pharmacology, and parasitology colleagues who may feel overlooked.

Tag E. Mansour
Stanford, California
July, 2001

Acknowledgments

Some of the research from my laboratory discussed in this book includes work done with research assistants, graduate students, postdoctoral fellows, and other colleagues to whom I am indebted for their interest in parasites and their dedication. I also acknowledge with gratitude the most current information I received from many colleagues outside Stanford who drew my attention to exciting developments in their own fields of research. My special thanks to Professor Miklós Müller for his critique of the first draft of Chapter 6 on Amitochondrial Protists. While writing this book I have been very fortunate in using Stanford's Lane Medical Library and Swain Chemical Library. I am indebted to the staff of both libraries for their expertise and their constant support.

1

The Search for Antiparasitic Agents

Early Beginnings of Chemotherapy

Historical records from almost every culture have examples of successful treatment of different ailments by potions, mainly from vegetable sources and occasionally from animals. A good example of an antiparasitic agent is cinchona bark (quinine), which was used against malaria by the early Peruvians and introduced to the Western World by the Jesuits. Extracts of male fern and of santonin obtained from *Artemisia maritima* var. *anthelmintica* were held in high regard as anthelmintics by Theophratus (370–285 B.C.) and Galen (A.D. 130–201) and were also recommended as effective anthelmintics until the 1940s and 1950s (Goodman & Gilman, 1955).

The beginning of chemotherapy as a science had to await the establishment of the germ theory of infection by Louis Pasteur more than a century ago. Rather than accepting the concept of spontaneous generation of infection, Pasteur suggested that diseases are caused by microbes that are transmitted among humans. Pasteur's inspiration for this theory came during his early studies on fermentation for the wine industry in France. Pasteur's contributions influenced bacteriology, immunology, the study of antibiotics, sterilization, epidemiology, and public health. This remarkable era was followed by the discovery and identification of many of the infectious microorganisms and methods for their culture in the laboratory and for infecting experimental animals. The study of parasites was initiated at the time of Sir Patrick Manson (1844–1922), who established the first institute for studying and teaching about tropical diseases: The London School of Hygiene and Tropical Medicine. Manson was a prominent researcher who made the seminal identification of the life cycle of the filarial nematode *Wuchereria bancrofti* in humans and the mosquito vector. He emphasized the importance of studies on tropical diseases and parasitology. The early part of the twentieth century saw the identification of many of the main parasites that afflict humans in the tropics. These included trypanosomiasis (sleeping sickness), leishmaniasis (kala azar), lymphatic filariasis, and malaria.

Paul Ehrlich and the Principle of Selective Toxicity

The father of chemotherapy was Paul Ehrlich (1854–1915), a physician initially interested in immunology. He won the Nobel Prize in 1908 for his work in the field of immunology following Emil Behring (1901), Ronald Ross (1902), and Robert Koch (1905). Ehrlich was fascinated by the selectivity of interaction between antigens and the antibodies that could be produced to them by animals. In the 1870s and 1880s, while using different dyes to stain tissues and parasites, he saw a similar kind of selectivity. Some of these dyes could bind to certain cells and even to certain parts of the cell whereas other cells remained free of dye. He was attracted to the idea that selectivity of binding of chemical agents that are toxic to parasites but not to human cells could be the basis of identifying effective chemotherapeutic agents. These observations in his laboratory inspired his thoughts about the basic principles of chemotherapy. It also proved Pasteur's famous dictum "Chance only favors the prepared mind."

During the first decade of the twentieth century there was a great interest among European countries in exploiting the natural resources from their African colonies. A way had to be found to protect the European colonizers and the native workers they employed from endemic diseases in the tropics. Attempts to use their knowledge of immunology were not successful in protecting people against parasitic infections. In addition, very little was known about the biology of these parasites. There was a great need for chemical agents to treat these infections.

Germany, at that time, had one of the largest and most modern chemical industries in Europe, particularly for the manufacture of dyes. Guided by his early observation that some dyes have selective affinity to certain cells and the availability of thousands of dyes from the neighboring Höchst laboratories Ehrlich embarked on a program to identify dyes that would have selective affinity to microorganisms responsible for different infections. He started with methylene blue, which showed some effects against malaria but also had a specific affinity to nerve cells and was therefore too toxic to humans. Hundreds of dyes were examined against both malaria and trypanosome infections in experimental animals without success.

Paul Ehrlich's new strategy was based on rational screening for identifying new chemotherapeutic agents. One should try to find chemicals that have been proven to be selectively toxic to the microorganisms but not to the human hosts. Ehrlich's thinking was greatly influenced by the specificity of antibody formation in response to antigens from infectious microorganisms. He perceived an analogy between the process of antigen/antibody interaction and the selective toxicity of chemical agents against parasites. He advocated the search for a toxic chemical agent that can selectively bind to a receptor in the parasite but not to one in the host. In an address to the 17th International Congress of Medicine in England (Ehrlich, 1913) he presented his ideas about the rational approach to chemotherapy. His first principle was "If the law is true in chemistry that *Corpora non agunt nisi liquida* (substances are effective only in the fluid state), then for chemotherapy the principle is true that *Corpora non agunt nisi fixata* (only attached, anchored substances are effective)." A metaphoric term "the magic bullet" was given to the ideal chemotherapeutic agent, a drug that disables the "germ" but not the host cells. Ehrlich envisioned a "*therapia sterilisans magna* ... that by means of one or at most two injections the body is freed from the parasites." Chemotherapy is a term that was invented by Paul Ehrlich and was defined as the use of chemical agents to "injure" an invading organism without harming the host.

In 1906, after his disappointing work with dyes, Ehrlich shifted his attention to the spirochetes that cause syphilis (which was widespread in Europe at that time). Mercury was the only known treatment, but this metal did not meet Ehrlich's conception of *therapia sterilisans magna*. Ehrlich studied arsenicals for his first research on identifying new, effective chemotherapeutic agents against syphilis. At that time Atoxyl (sodium arsanilate) was being used as an antiprotozoal agent, but because of toxicity, its therapeutic usage was discontinued. Several hundred arsanilate derivatives were examined against an infection of a spirochete-like organism, *Spirillum morsus muris*, in laboratory animals. Compound 606 (diaminodioxyarsenobenzol) was successful in treating *Spirilla* infections in chickens, rabbits, rats, and mice because *Spirilla* readily absorbed the drug, killing the bacterial cells (Baumler, 1965). Subsequently, Compound 606 was used with a dramatic effect in patients suffering from syphilis. The human cells did not appear to take up the drug. An improved preparation of Compound 606 (neosalvarsan, Na diamino dihydroxyarsenobenzene methanol sulfoxylate), which was more soluble and thus more suitable for injection, was subsequently discovered. Within five years the incidence of syphilis was greatly reduced in many European countries. This was a triumphant success.

Rational Discovery of Antiparasitic Agents

Subsequent to the work of Paul Ehrlich, a strategy was gradually established to search for antiparasitic agents on the basis of identifying drugs that are selectively toxic to the parasite with either no or minor effect on the host. This is now referred to as rational drug design. Because of its intellectual appeal the rational approach has been espoused predominantly by scientists in academic medicine. With the advances in biochemistry, molecular biology, and the techniques for synthesizing new chemical agents, significant progress has been made toward achieving Paul Ehrlich's goals. The ideal chemotherapeutic targets are those that are unique to the parasite. A few examples, which will be more fully discussed later in this book, can best explain this approach.

In trypanosomes regulation of an intracellular reducing milieu is controlled by an enzyme that is quite different from that in mammals. Glutathione reductase catalyzes the transfer of electrons from NADPH to oxidized glutathione in mammals. In trypanosomes, instead of glutathione, a peptide polyamine conjugate of glutathione and spermidine (named trypanothione) is used. Reduction of trypanothione is catalyzed by trypanothione reductase (Fairlamb & Cerami, 1992). Because of its uniqueness, trypanothione reductase is a favorite target for screening prospective trypanosomicidal agents. Melarsoprol (a trivalent organic arsenical) and nifurtimox (a nitrofuran derivative) are known to be inhibitors of trypanothione metabolism and are used as trypanosomicidal agents. The opportunity was, therefore, open for identifying new and more effective antitrypanosomal agents using the reductase system as a drug target (see Chapter 5).

Another example of a target is dihydropteroate synthase (DHPS). This enzyme catalyzes the first reaction involved in the synthesis of folic acid. Several protozoal parasites including *Plasmodium, Toxoplasma*, and *Eimeria* cannot absorb folic acid from the host and they depend on their own *de novo* synthesis of folic acid. DHPS is considered to be a good target for antiparasitic agents in these organisms (Wang, 1997). The enzyme is inhibited by sulfonamides or pyrimethamine. These compounds have antiparasitic effects and are used therapeutically in patients (see Chapters 4 and 5).

Biology of Parasites

In the past decade there have been great advances in our understanding of the biology of parasites. Recent discoveries regarding the physiology and biochemistry of protozoal and helminthic parasites have elucidated many of the prospective targets that are unique to each parasite. There is currently more basic biological information available than has been sufficiently exploited for chemotherapy of parasites. The rational approach to the discovery of new antiparasitic agents is to identify as many of these prospective targets as possible and to aim the new chemical agents selectively against these targets.

Although parasites may be quite biologically diverse, there are examples of chemotherapeutic agents that are effective against more than one type of parasite. A compound that is effective against one species may very well be effective against other unrelated organisms. Sulfonamides, originally discovered for their effect against streptococci, were later found to be active against chicken malaria (an experimental infection) as well as chicken coccidiosis. Therefore, it is important to study the mechanism of action of these agents in species other than those where they are presently used. Scientists directed their attention to the physiology and biochemistry of these organisms to explain the mechanism of action of certain effective chemotherapeutic agents. Some of the best antiparasitic agents were recognized long before a modern understanding of the biology of these parasites was established. Elucidation of their mechanism of action usually uncovered targets that are amenable to further biological studies. Biologists who are interested in the discovery of antiparasitic agents should study what parasites need from their hosts.

Biochemical Adaptation of Parasites to the Host

If it is assumed that parasites were derived from free-living ancestors then they must have been subjected to the process of natural selection within the mammalian host. This evolutionary process must have favored their ability to metabolize food from the host for energy and for reproduction. They must also have developed ingenious biochemical systems for self-maintenance within the unfriendly environment of the host. These evolutionary processes result in the selection of metabolic pathways and enzymes that are adapted for the survival of the parasites in their hosts. Evolution of the parasite had to keep pace with evolution of the host. Since internal parasites were not subjected to the same environmental factors that influence their hosts, the parasites have resorted, in many cases, to strategies that are different from those of the host. A rational approach to selection of antiparasitic agents must include identification of biochemical and molecular differences between the parasite and the host. Such differences could be manifested in the form of a biochemical pathway or an enzyme that is unique to the parasite. This would represent the ideal target against which antiparasitic agents should be designed. Evolutionary adaptation of the parasite to the host may not have involved drastic changes. The parasite may use metabolic pathways that are similar to those of the host, but it may have adapted its own enzymes to have different structures or different kinetic properties. Differences between the parasite and host enzymes may include differential sensitivity to inhibitors. Knowledge of the biochemical and molecular differences between parasite and host and of rate-limiting points in the parasite metabolism are both necessary for successful selection of targets.

Parasites are well adapted for survival in the host's environment. Here they satisfy all their needs, while usually providing no benefit to the host. The result of this one-sided dependency is that it is not in the parasites' best interests to kill the host and in many cases parasites are able to survive for a long time. Such habituation may deprive the host of essential nutrients and usually exposes the host's tissues to the parasites' metabolic and degradative products, causing pathological changes in and damage to the host. Obviously, remedies against parasites should subsequently include treatment of the host for clinical and pathological changes induced by the parasites.

Drug Receptor Selectivity

The term "drug receptor" has been loosely used to designate the primary site that selectively binds a drug. The more restrictive use of the term is: a macromolecule that has as one of its main functions the binding of a hormone or a neurotransmitter that is naturally present in the body (e.g., acetylcholine for nicotinic receptors). Throughout this book I shall be using the restrictive usage of the term "drug target" to designate the macromolecule that binds a drug selectively. Through the use of modern biological techniques our knowledge of antiparasitic agents has been sufficiently advanced to identify some targets that are critical for the survival of parasitic organisms. In most of these cases several chemical agents that interact with these targets have been screened by different mechanisms. There is still, however, a need to establish the stereospecificity of chemical agents for obtaining optimal biological effect against the parasite. It is necessary to establish an assay procedure that is accurate and simple. This could be achieved by isolating the target and measuring the binding of the candidate chemicals to the target or by examining the biological effect of the chemical on enzyme activity. Mathematical calculations of structure–effect data for chemical analogs are important for deducing the relationship between functional chemical constituents of the drug and the biological effect on the target. Information regarding the optimal size and shape of the drug, the hydrophilic/lipophilic ratio, and the reactive functional groups that determine the selectivity of the binding to the parasite target are all crucial in selecting for potent antiparasitic agents. The nature of the target should be investigated. A requirement for certain reactive groups in the prospective drug that favor binding may suggest the presence of complementary groups in the target molecule.

Metabolic Pathways as Targets

Human as well as parasite cells have hundreds of metabolic pathways that are vital for their normal function. There is always a need to study mammalian pathways and compare them with those of the parasite. Each pathway has a number of enzyme reactions that catalyze different steps in the pathway. Enzyme activity at every step is regulated to ensure that the final product of the pathway provides for the needs of the cell. Metabolic pathways are interconnected to others. Each enzyme reaction in a pathway is governed by the law of mass action. In each pathway one or more enzymes are considered to be rate limiting or pacemaker enzymes. These enzymes usually are endowed with allosteric control mechanisms. Activity of these regulatory enzymes is subject to feedback inhibition or activation by one or more of the substrates in the pathway. Regulation of these key enzymes is critical for coupling the activity of the pathway to the metabolic needs of the cell. The Embden–Meyerhof glycolytic pathway is a good example of regulation of a multienzyme system. This pathway has eleven enzymes to convert glucose to lactic acid with the net production of two molecules of ATP. Three of these enzymes are considered to be rate limiting: hexokinase, phosphofructokinase, and pyruvic kinase. If cellular levels of ATP generated by glycolysis are high, ATP inhibits the three rate-limiting enzymes, particularly phosphofructokinase. Higher levels of AMP and ADP favor the "deinhibition" of these enzymes and thus favor increasing glycolysis and ATP production. Rate-limiting enzymes such as phosphofructokinase are usually good enzyme targets to inhibit the vitally important glycolytic pathway (Mansour, 1979). In a multienzyme pathway inhibition of two enzymes by two different drugs can have a more inhibitory effect on the metabolic pathway than the use of either drug alone. An example of such synergistic effects can be seen in the synthesis of tetrahydrofolic acid. In the treatment of malaria the use of a sulfonamide that inhibits dihydropteroate synthetase together with trimethoprim that inhibits dihydrofolic acid reductase has a greater effect than using either drug alone (see Chapter 4).

In the past, studies on the nature of enzymes have focused mainly on their properties in the purified form. Although kinetics and other biophysical information about isolated purified enzymes are essential in deducing their ultimate physiological role, very often regulation of these enzymes within the constraints of activity of other enzymes in the metabolic pathway cannot be predicted. It is now understood that in a multienzyme system each enzyme is influenced by the activity of preceding and following enzymes in the pathway as well as extrinsic signaling systems outside the pathway. The regulation of these enzymes is made more complicated by the fact that some components of the multienzyme system are regulated by collateral signaling systems such as protein kinases or transmembrane transport of substrates for the metabolic pathway. We now have a better understanding of such regulation in many mammalian pathways, such as the glycolytic pathway in muscle or liver cells.

Once an enzyme is found to catalyze a reaction that limits the rate of an essential metabolic pathway, more information will be needed about its properties and chemistry. The next essential step for these studies is isolation and purification of the enzyme. Studies of the properties of an enzyme in crude extracts very frequently give misleading information. Enzyme specificity, regulation, kinetics, and other physical parameters cannot be studied with certainty using impure enzyme preparations. Detailed information regarding the chemistry of the enzyme, particularly the active site, can be obtained by cloning the enzyme and determining the amino acid sequence (see Chapter 2). In addition, the cloned enzyme can be expressed in appropriate bacterial or eukaryotic cells to provide ample amounts of material for further studies as well as for screening inhibitors.

Empirical Screening of Antiparasitic Agents

Most of our currently used antiparasitic drugs have been identified as a result of random screening of a series of chemicals that are related to compounds with recognized therapeutic value. This is referred to as the empirical approach to drug discovery. A successful example of this approach was the vast research program of the United States Army at Walter Reed Institute to screen for chemicals with potential antimalarial effects, focusing particularly on malaria strains resistant to chloroquine or pyrimethamine or both (Schmidt et al., 1978). A special group of chemicals considered were the 4-quinolinemethanols. This group was chosen because quinine, the first naturally occurring antimalarial agent discovered, belongs to the 4-quinolinemethanol group. Moreover, quinine was an effective antimalarial agent against multidrug-resistant strains of P. falciparum. Of the thousands of 4-quinolinemethanol derivatives tested in mice infected with Plasmodium berghei (mouse malaria), 300 quinolinemethanol derivatives were chosen. In considering chemotherapeutic data based on the use of mice, the question always arises whether these compounds would be equally effective against the human malarial species P. falciparum. In a pilot experiment the most potent 4-quinolinemethanol derivatives were tested against resistant P. falciparum in owl monkeys and the best eight compounds were selected. These were tested for their antimalarial properties and tolerability in human volunteers. Six of the eight compounds tested were found to be less active than quinine. Compound WE-142,490 (subsequently given the name mefloquine) was among the most effective agents against resistant *P. falciparum*. It was effective when given as a single dose and it was also effective against *Plasmodium vivax*.

Use of Animal Models in Screening

The advantages and disadvantages of the empirical screening procedure described above depend in part on the number of animals needed. In the case of *P. falciparum*, the need for testing in owl monkeys was a disadvantage because of the limitation on the number of available animals and the high cost of maintenance. The human malarias are very specific about the host in which they can live and thus cannot be transmitted to small laboratory animals. Furthermore, antimalarial agents that are effective against mouse or bird malaria do not necessarily have chemotherapeutic value in humans. In the case of many other parasites, rodents or birds have been of use only for a first screening before human subject trials. In addition to high cost, using experimental animals for empirical screening of a large number of compounds entails adherence to extensive government regulations. However, examining the effects of drugs against the parasites in their natural animal habitat has several advantages. One advantage in using laboratory animals for drug screening is that information on the pharmacokinetic and toxic properties of the drug can be obtained earlier (Hudson, 1994).

Use of *in vitro* Parasite Cultures for Screening

In vitro cultures have been extensively used for screening chemotherapeutic agents against bacteria. This technique is less costly than animal screening but depends on the ease of establishing laboratory cultures. An additional advantage is the opportunity to genetically manipulate most of these organisms. Bacterial cultures can also be used for vaccine development. In the case of most of the protozoal and helminth parasites it has not always been possible to establish continuous *in vitro* cultures. The relationship between the parasite and the host is very intricate and disruptions of the host–parasite relationship *in vitro* leads to gradual death of the parasite. Studies on the metabolism or physiology of movement were carried out for only a short period of time after removing the parasite from the host. Attempts are always made to provide the parasite with everything the investigator can think of, but it is often difficult to keep a continuous culture of the parasite. One obvious disadvantage of *in vitro* tests is that some compounds being tested (pro-drugs) may have to be metabolized by the host to an active form and therefore would not be recognized by *in vitro* screens.

In many cases maintaining the complete life cycle of a human parasite in the laboratory has not yet been possible. For many years numerous scientists tried to establish the erythrocytic stage of *P. falciparum*, in the laboratory. Trager achieved a breakthrough in maintaining this stage of the human malaria parasite in continuous *in vitro* culture (Trager & Jensen, 1976). The procedure involved using an appropriate medium over a layer of human red cells maintained in a gas phase of 3–5% CO_2 and less than 21% oxygen. Media preparation and the correct gas content are now so simple that the practice of culturing human malaria has been established all over the world even in relatively primitive laboratory conditions. This procedure has enabled scientists in many countries to learn more about the biochemistry and cell biology of the parasite and about chemotherapy and drug resistance. The *in vitro* culture techniques have been used to study the mechanism of action of antimalarial agents and to screen for antimalarial agents.

In the case of *Toxoplasma gondii*, one of the opportunistic infections that afflict some AIDS patients, the tachyzoite stage of the parasites can be harvested from the peritoneal cavities of infected mice and cultured *in vitro* (Remington, Krahenbuhl, & Mendenhall, 1972). Replication of *T. gondii* can be measured by determining the incorporation of [^{3}H] uracil by the tachyzoites, while the mammalian host cells show no significant incorporation of uracil (Pfefferkorn & Pfefferkorn, 1977). Chemical agents being tested are added to the medium after the *T. gondii* have infected the host cells.

With *Trypanosoma cruzi* (the causative agent of Chagas' disease) HeLa cells are used as the host. They are grown for two days before being infected with the trypomastigotes. This stage is flagellated and nondividing. The trypomastigotes are transformed to the amastigote stage after invading the mammalian cells. The percentage of HeLa cells that have one or more amastigotes (intracellular stage) is recorded every day for 5–7 days (Nakajima-Shimada, Hirota, & Aoki, 1996). A similar procedure for *in vitro* drug screening against *T. brucei rhodesiense* was established using human HL-60 feed-layer cells. The number of trypanosomes and of HL-60 cells are counted using a hemocytometer and concentrations of drugs causing 50% growth inhibition are determined.

Leishmania mexicana, a dermatological protozoal infection, has, like trypanosomes, two forms: the flagellated promastigotes, which live in the invertebrate vector, and the nonflagellated amastigotes, which multiply in the vertebrate host. A medium developed by Pan was found to be optimal for *in vitro* culture and rapid transformation of the *Leishmania* promastigotes to amastigotes (Pan, 1984). These cultures can be used for drug screening.

Similar *in vitro* culture experiments were carried out on the protozoan parasites *Giardia duodenalis*, *Trichomonas vaginalis*, and *Entamoeba histolytica*. All cause intestinal infections in humans. A specific culture medium has been developed for each parasite. The trophozoites are harvested for the drug test experiments when they are in the mid-logarithmic growth phase. Parasite viability is evaluated every day by observing motile parasites. Minimum lethal concentration of the drugs being tested are determined (Upcroft *et al.*, 1999).

It has been difficult to establish continuous cultures of parasitic helminths because of their complex life cycle. In addition to some of them having an invertebrate as intermediate host, many of these parasites have one or more free-living stages. Some of these helminths live in the gut of the host, which creates a serious problem of contamination of the cultures with a variety of gut microorganisms. Because of these considerations *in vitro* drug screening tests are usually limited to short time incubations immediately after the parasites are detached from the host. Most of the parasitic helminths can be kept in the presence of antibiotics under physiological conditions for at least 24 hours. *Ascaris suum*, because of a thick cuticle and large reserves of glycogen, can be kept for several days in a simple saline medium (Bueding, Saz, & Farrow, 1959). *Fasciola hepatica* can also be maintained in good physiological condition for at least 24 hours in a special saline medium containing glucose and antibiotics.

In the case of *Schistosoma mansoni* there has been some success in culturing adult schistosomes from cercariae that were transformed to schistosomula. However, only a few of the mature adults produce eggs, which in most cases are abnormal, indicating that the *in vitro* development was not normal (Basch & Rhine, 1983). For routine *in vitro* testing of drugs adult schistosomes can be removed from infected mice and maintained in culture media (Bueding, 1950). The effects of tested drugs can be determined by survival of the parasites versus controls for a limited time.

In vitro culture of parasitic cestodes has enjoyed greater success. Many of their intermediate stages live in fish or vertebrate muscle and can be isolated under sterile conditions. The entire life cycle of *Hymenolepis diminuta* has been cultured *in vitro* (Smyth, 1990). This has been especially valuable for studies on the developmental biology of the parasites as well as for screening chemical agents against cestodes.

Among nematodes that live in human tissues, filarial worms are of great medical importance. Recently, cultivation of the third-stage larvae of *Brugia malayi* and their transformation to the young adult stage has been reported (Falcone *et al.*, 1995). Of the larvae cultured 17% reached young adulthood after 37 days. This is an important contribution to studies on development of this filarial parasite. These cultures could also be used for *in vitro* screening of chemical agents. *Brugia* could be used as a representative of other human filarial parasites such as *Wuchereria bancrofti* (which causes elephantiasis) and *Onchocerca volvulus* (which causes onchocerciasis or river blindness).

Designing Chemicals for Selective Targets

The grand revolution in biochemistry during the last half of the twentieth century opened new fields of research for the development of selective antiparasitic agents. Much can be learned from drug development strategies for anticancer and antimicrobial drugs. A classical example is penicillin, the champion of antibacterial agents. The target of penicillin is the cell wall of *Streptococci*, a structure that is not present in host cells. Enzymes that are present in both parasite and host can also be amenable to inhibition by proper design of chemical agents to selectively inhibit the parasite enzyme. The ideal inhibitor would be a chemical that reacts with the target enzyme better than the natural substrate of that enzyme.

Based on the chemical structure of enzyme substrate and possible modifiers, different tentative inhibitors can be selected and tested on enzyme activity. Once a promising inhibitor is chosen, analogs of the molecule should be synthesized with the aim of improving the inhibitory potency of the compound. A measure of the inhibitory effect of the compound on mammalian as well as parasite cells is essential in determining the selectivity of the inhibitors. Chemical changes to improve its selectivity may include addition of new hydrophobic groups to make the molecule more able to penetrate the cells or making the inhibitor more robust by modifying the molecule by cyclization. For reviews of the general principles of design of inhibitors see Stark & Bartlett (1983) or Smith (1988).

Reversible inhibitors bind to the target enzyme physicochemically, either competitively or noncompetitively depending on their relationship to the enzyme–substrate reaction. An example is the use of sulfonamides, the therapeutic efficacy of which can be overcome by p-aminobenzoic acid (see Chapter 4).

An example of an irreversible inhibitor is the effect of metronidazole on *Entamoeba histolytica* (see Chapter 6). Transition state analogs or intermediate analogs that have high affinity to the enzyme's substrate site can also be drugs. In many cases the structure of these transition state analogs can be predicted from the mechanism of the enzyme reactions and/or by NMR analysis. Subversive (suicidal) inhibitors (mechanism-based inhibitors) are false substrates for an enzyme, which the enzyme itself converts to an inhibitor. These inhibitors interact preferentially with the target enzyme at the active site, but the enzyme reaction is not completed and the enzyme is inactivated. Allopurinol is a hypoxanthine analog that acts as a false substrate of the purine phosphoribosyl transferase of some protozoal parasites that cannot synthesize purine *de novo*. Allopurinol is not toxic to humans because they can synthesize purines *de novo* (see Chapter 5).

In the case of a target enzyme that has two substrates, inhibitors could be designed as a single molecule that can recognize both sites. Such an inhibitor would have a higher order of specificity and is usually referred to as a multisubstrate analog inhibitor (MAI) (Broom, 1989). A classical example of MAI is N-(phosphonoacetyl)-L-aspartate (PALA), an inhibitor of aspartate transcarbamoylase. PALA binds to the enzyme three orders of magnitude more tightly than the natural substrate, carbamoyl phosphate. PALA has been shown to have antiproliferative effects in mice (Yoshida, Stark, & Hoogenraad, 1974) and against B16 melanoma and Lewis lung carcinoma (Johnson *et al.*, 1976) but has not been used for chemotherapy in humans.

Relationship between Chemical Structure and Biological Activity of Inhibitors

The accepted procedure for studying the relationship between chemical structure and biological activity is to choose a compound that severely inhibits the target enzyme. However, in many cases the drug target has not been isolated and one can only measure the biological effect of the drug. Frequently, there are several chemical relations between the target and the final biological effect. Because the biological effect is always triggered by the initial interaction between the drug and the target, the biological response to the compound is accepted as a measure of drug–enzyme interaction. It is customary to test a series of chemicals that are related to the original drug in an attempt to improve its antiparasitic potency and to reduce the toxic effects against host cells. Chemical modification of the drug molecule could include changes in size, shape, or side chains. After trying many compounds, one can deduce a rough idea of the three-dimensional structure of the drug target. When the enzyme is isolated and the chemical structures of inhibitors are known, more accurate mapping of the target molecule using computer graphics can be undertaken.

Quantitative Structure Activity Relationship

A methodology for drug design, known as Quantitative Structure Activity Relationship (QSAR), has been developed to quantitate the relationship between chemical structure of a compound and its biological activity. A description and an analysis of the methods have been written by James (1988). In QSAR different chemical substituents that modify the structure of a compound are assigned certain parameters. These give a measure of the contribution of the chemical groups to the lead compound used. The biological activity of different derivatives are determined and compared to the selective parameters of their chemical substituent groups. The structures of the most promising derivatives can be predicted mathematically. Electronic parameters that determine the effect of interactions between the chemical and the binding site of the target are important. A parameter for the steric substituent is also commonly used. This gives a measure of the bulkiness of the group and determines the closeness of contact between the chemical and the receptor site. In addition, parameters related to solubility of the compound are used in determining the relationship between structure and activity. The mathematical concepts of QSAR have enabled researchers in drug design to use a more quantitative approach.

Combinatorial Technology in Drug Discovery

The most radical new development for selecting chemotherapeutic agents involves a technique known as combinatorial chemistry. Synthetic chemists have had the tradition for more than a century of going to elaborately equipped laboratories to synthesize individual derivatives of lead compounds. No matter how long it took there was a professional pride in synthesizing, purifying, and confirming the structure and properties of individually prepared chemicals. During the past ten years the preparation of combinatorial libraries has drastically changed drug discovery and development. This new technology generates molecular diversity through the use of small reactive molecules, referred to as the building blocks (Gallop et al., 1994). These react with a constant backbone chemical, producing a variety of different individual compounds. In an ideal combinatorial procedure the number of derivatives is determined by the number of building blocks for each step and the number of synthetic schemes. The procedure can be exploited to build hundreds of thousands of derivatives. In dealing with a large number of chemicals, perhaps millions of derivatives, automated techniques can be robotized. Historically, the procedure was initiated for the development of peptide libraries. The availability of a variety of amino acid building blocks with well-established peptide chemistry contributed to the versatility of the new technology in making highly diverse peptide libraries.

Combinatorial technology has been extended to the preparation of oligonucleotides and more recently to the generation of small molecule libraries by combinatorial organic synthesis (Gordon et al., 1994). There are a variety of combinatorial assays used to identify and characterize synthesized ligands. In the case of immobilized target receptors the compounds are assayed by affinity purification techniques. If the ligands are tethered (to pins, beads, or chips, etc.) the identity of the ligands are either determined directly by mass spectroscopy or, in the case of peptides, by sequencing. Libraries of soluble chemicals are tested individually in competition binding assays or enzyme assays or by determining their effects in a cell-based bioassay. Attention must always be paid to discriminate between specific and nonspecific binding (Gordon *et al.*, 1994; Thompson, Klein, & Geary, 1996).

Combinatorial search for new antiparasitic agents does not fit well with the empirical or the rational screening models. It is based on random screening of a vast number of randomly synthesized chemical agents. For the incredible numbers of new chemicals that can be screened by the combinatorial techniques it is fast and cheap once it is established in the laboratory. The expectation is that the results of genomic analyses of several model protozoa and helminth parasites will be available in the next three to five years. A draft of the human genome is now available. Analysis of all these genomic data should place investigators in a good position to identify enzymes and receptors that are unique to the parasites. Molecular biological technology should make it possible to express these genes and use the pure proteins (see Chapter 2) in a combinatorial screen to search for new antiparasitic agents among thousands of chemical derivatives.

The power of combinatorial methods has been shown to be successful in identifying new drugs. This has recently been illustrated in the search for new inhibitors of aspartyl plasmepsin II, one of the proteolytic enzymes of hemoglobin in *P. falciparum* (Carroll & Orlowski, 1998). The malaria depends on human hemoglobin as the main source of food. Inhibitors of plasmepsin II prevent hemoglobin degradation and kill the parasites in culture. The enzyme is therefore an excellent target for identifying antimalarial agents. A combinatorial chemistry library was established using ECLiPS (Encoded Combinatorial Library Polymeric Support) technology (Burbaum *et al.*, 1995). Pure plasmepsin II was used with chemical data from several preliminary large-scale combinatorial chemistry libraries. A synthetic path was used to generate the individual compounds, which were identified by special tags and eventually by NMR and mass spectrometry. Out of a total of 32,000 compounds a few had a K_i value of $\sim$100 nM and appear to have selective inhibitory effects against plasmepsin II of *P. falciparum*.

The use of combinatorial RNA that binds to the surface of live African trypanosomes to isolate high-affinity ligands was reported by Homann and Goringer (1999). They were trying to selectively tag the membrane surface proteins of trypanosomes without the interference of the expressed variant surface glycoproteins (VSGs). Short RNA ligands (aptamers) were used to bind to the membrane proteins of live trypanosomes. Because of their small size the aptamers can glide in between VSGs, avoiding their physical barrier, and then bind specifically and with high affinity to cell-membrane proteins. These could include receptor complexes, transporters, and other invariant surface glycoproteins. Using this new technology Homann and Goringer were able to identify a 42-kDa protein located within the flagellar pocket of the parasite. The protein was found to be a subunit of transferrin, an iron transfer protein. The same procedure can be used to identify other enzymes or transporters as targets for prospective antiparasitic agents. The possibility has been raised that binding of the RNA to the flagellar pocket (the main site for endocytosis and exocytosis) in the parasite may have identified a novel target for antitrypanosomal agents.

The most important development in chemotherapy of parasites during the end of the twentieth century has been the acceptance of the rational approach to chemotherapy. Current explosive developments in biochemistry, molecular biology, and physiology of parasites provide a basis for identifying selective targets in the parasites. There is general consensus that antiparasitic agents should be designed to take advantage of the biochemical differences between parasite and host. Rational design of antiparasitic agents against selective targets is facilitated by the new knowledge of combinatorial chemistry. In the next chapter the advances in biophysical, genomic, and proteomic techniques that have revolutionized this area are discussed.

REFERENCES

Basch, P. F. & Rhine, W. D. (1983). *Schistosoma mansoni*: Reproductive potential of male and female worms cultured in vitro. *J Parasitol, 69*(3), 567–569.

Baumler, E. (1965). *In Search of the Magic Bullet*. London: Thames and Hudson.

Broom, A. D. (1989). Rational design of enzyme inhibitors: Multisubstrate analogue inhibitors. *J Med Chem, 32*(1), 2–7.

Bueding, E. (1950). Carbohydrate metabolism of *Schistosoma mansoni*. *J Gen Physiol, 33*, 475–495.

Bueding, E., Saz, H. & Farrow, G. (1959). The effect of piperazine on succinat production by *Ascaris lumbricoides*. *Brit J Pharmacol, 14*, 497–500.

Burbaum, J. J., Ohlmeyer, M. H., Reader, J. C., Henderson, I., Dillard, L. W., Li, G., Randle, T. L., Sigal, N. H., Chelsky, D. & Baldwin, J. J. (1995). A paradigm for drug discovery employing encoded combinatorial libraries. *Proc Natl Acad Sci USA, 92*(13), 6027–6031.

Carroll, C. D. & Orlowski, M. (1998). Screening aspartyl proteases with combinatorial libraries. *Adv Exp Med Biol, 436*, 375–380.

Ehrlich, P. (1913). Chemotherapeutics: Scientific princples, methods, and results. *Lancet, 2*, August 16, 445–451.

Fairlamb, A. H. & Cerami, A. (1992). Metabolism and functions of trypanothione in the Kinetoplastida. *Annu Rev Microbiol, 46,* 695–729.

Falcone, F. H., Zahner, H., Schlaak, M. & Haas, H. (1995). In vitro cultivation of third-stage larvae of *Brugia malayi* to the young adult stage. *Trop Med Parasitol, 46*(4), 230–234.

Gallop, M. A., Barrett, R. W., Dower, W. J., Fodor, S. P. & Gordon, E. M. (1994). Applications of combinatorial technologies to drug discovery. 1. Background and peptide combinatorial libraries. *J Med Chem, 37*(9), 1233–1251.

Goodman, L. & Gilman, A. (1955). *The Pharmacological Basis of Therapeutics* (2nd ed.). New York: MacMillan.

Gordon, E. M., Barrett, R. W., Dower, W. J., Fodor, S. P. & Gallop, M. A. (1994). Applications of combinatorial technologies to drug discovery. 2. Combinatorial organic synthesis, library screening strategies, and future directions. *J Med Chem, 37*(10), 1385–1401.

Homann, M. & Goringer, H. U. (1999). Combinatorial selection of high affinity RNA ligands to live African trypanosomes. *Nucleic Acids Res, 27*(9), 2006–2014.

Hudson, A. (1994). The contribution of empiricism to antiparasite drug discovery. *Parasitol Today, 10,* 387–389.

James, K. (1988). Quantitative structure–activity relationships and drug design. In H. Smith (Ed.), *Smith and Williams Introduction to the Principles of Drug Design* (2nd ed., pp. 240–264). London: Butterworth & Co.

Johnson, R. K., Inouye, T., Goldin, A. & Stark, G. R. (1976). Antitumor activity of N-(phosphonacetyl)-L-aspartic acid, a transition-state inhibitor of aspartate transcarbamylase. *Cancer Res, 36*(8), 2720–2725.

Mansour, T. E. (1979). Chemotherapy of parasitic worms: New biochemical strategies. *Science, 205*(4405), 462–469.

Nakajima-Shimada, J., Hirota, Y. & Aoki, T. (1996). Inhibition of *Trypanosoma cruzi* growth in mammalian cells by purine and pyrimidine analogs. *Antimicrob Agents Chemother, 40*(11), 2455–2458.

Pan, A. A. (1984). Leishmania mexicana: Serial cultivation of intracellular stages in a cell-free medium. *Exp Parasitol, 58*(1), 72–80.

Pfefferkorn, E. R. & Pfefferkorn, L. C. (1977). Specific labeling of intracellular *Toxoplasma gondii* with uracil. *J Protozool, 24*(3), 449–453.

Remington, J. S., Krahenbuhl, J. L. & Mendenhall, J. W. (1972). A role for activated macrophages in resistance to infection with *Toxoplasma. Infect Immun, 6*(5), 829–834.

Schmidt, L. H., Crosby, R., Rasco, J. & Vaughan, D. (1978). Antimalarial activities of various 4-quinolonemethanols with special attention to WR-142,490 (mefloquine). *Antimicrob Agents Chemother, 13*(6), 1011–1030.

Smith, H. (1988). *Introduction to the Principles of Drug Design* (2nd ed.). London: Wright.

Smyth, J. (1990). *In vitro Cultivation of Parasitic Helminths.* Boca Raton, FL: CRC Press.

Stark, G. R. & Bartlett, P. A. (1983). Design and use of potent, specific enzyme inhibitors. *Pharmacol Ther, 23*(1), 45–78.

Thompson, D. P., Klein, R. D. & Geary, T. G. (1996). Prospects for rational approaches to anthelmintic discovery. *Parasitology, 113* (Suppl), S217–238.

Trager, W. & Jensen, J. B. (1976). Human malaria parasites in continuous culture. *Science, 193*(4254), 673–675.

Upcroft, J. A., Campbell, R. W., Benakli, K., Upcroft, P. & Vanelle, P. (1999). Efficacy of new 5-nitroimidazoles against metronidazole-susceptible and -resistant *Giardia*, *Trichomonas*, and *Entamoeba* spp. *Antimicrob Agents Chemother, 43*(1), 73–76.

Wang, C. C. (1997). Validating targets for antiparasite chemotherapy. *Parasitology, 114*(Suppl), S31–44.

Yoshida, T., Stark, G. R. & Hoogenraad, J. (1974). Inhibition by N-(phosphonacetyl)-L-aspartate of aspartate transcarbamylase activity and drug-induced cell proliferation in mice. *J Biol Chem, 249*(21), 6951–6955.

2

Biophysical, Genomic, and Proteomic Analysis of Drug Targets

Biophysical Techniques for Studying Drug Targets

X-Ray Crystallography

Once a selective parasite target has been identified the next step is to obtain its three-dimensional (3-D) structure. Crystallography remains the most desirable method for understanding the structure of the target, its binding sites, and the chemical groups involved in electron and steric interaction with substrates or inhibitors. The initial step toward achieving this goal is the preparation of large enough crystals of the protein. In the past this has been difficult because of the scarcity of parasite material as a source of the target protein. This is now a less serious handicap because the molecular biological techniques of cloning, polymerase chain reaction (PCR), and expression of target genes in bacterial or insect cells can be used. It is now possible to obtain highly purified recombinant proteins in milligram amounts. The protein is placed in a solution of ammonium sulfate, the concentration of which is gradually increased to avoid precipitation of an amorphous protein. Having patience and a "green thumb," which may come with practice, seem to be valuable elements for preparing crystals. Although the principles of X-ray crystallography were established more than fifty years ago, there have been many improvements in X-ray generators and detector systems. The development of synchrotron radiation sources has increased the accuracy and speed of analysis of crystals (Hunter, 1997). The development of computer analysis and graphics that can display different parts of the protein molecule on a monitor is very useful.

Trypanothione reductase (TR) is an enzyme found in trypanosomatids but not in mammals (Fairlamb & Cerami, 1985). TR and the corresponding mammalian enzyme, glutathione reductase, were subject to crystallographic analysis bound to their respective substrates (see Chapter 5).

Information about the 3-D structure of a target is not only useful for iden-tifying ligands for the original protein but can also be of value for designing inhibitors of structurally related targets. A serine protease from cercariae of *Schistosoma mansoni* was used to design a 3-D structural model taking advantage of a sequence similarity to a trypsin-like class of serine proteases (Cohen *et al.*, 1991). The same computer program was used to identify inhibitors of a *Plasmod-ium falciparum* cysteine protease that degrades hemoglobin, the main source of amino acids for the parasite. Using the computational model, Ring *et al.* (1993) found a nonpeptide inhibitor with a K_i of 3 μM that, when tried in *in vitro* malaria cultures, arrested growth and development of the parasites.

Nuclear Magnetic Resonance

Nuclear magnetic resonance (NMR) has recently been widely used in parasite research. It is particularly useful for noninvasive experiments on live intact para-sites for monitoring the effects of chemical agents on the metabolism of parasites. In addition, NMR has become an important technique for studies on molecular targets in parasites and their interaction with different ligands. This technique has the advantage over crystallography in that NMR examines the proteins and their ligands in solution. It is, therefore, possible to simultaneously detect and define motions at individual atomic sites. Therefore, whereas X-ray diffraction is powerful in identifying 3-D protein structures, NMR is useful for studying the dynamics of these molecules at atomic resolution. The development of high-field spectrometers and two-dimensional NMR provided significant new advances in NMR technology to search for selective inhibitors against targets in the parasite.

In a recent report by Li *et al.* (1999) transition-state analogs were identified as inhibitors of hypoxanthine-guanine phosphoribosyl transferases (HGPRTs) in the human malaria parasite *Plasmodium falciparum*. The salvage of hypoxanthine and/or xanthine through hypoxanthine-guanine-xanthine phosphoribosyltrans-ferase (HGXPRT) is necessary to satisfy the purine needs of the parasite. Li and his associates, using NMR, were able to show that HGXPRT bound to substrate gives downfield hydrogen-bonded protons. The chemical shifts move further downfield with bound analog inhibitors such as immucillin HP or immucillin GP. Immucillin-5-phosphates have been identified as transition-state analog in-hibitors of the human and malarial HGXPRTs. The equilibrium inhibition con-stant is three orders of magnitude greater than that of the physiological substrate (IMP). The discovery of this group of potent inhibitors is a major step toward the design of related analogs that would be more selective against the malarial HGXPRT. This enzyme is discussed more fully in relation to trypanosomes in Chapter 5.

Over the past twenty years NMR has been used as a tool for studying metabolic pathways in living parasites. Many of these pathways have proven to be good targets for antiparasitic agents. ^{31}P-NMR was used for studying the liver fluke *Fasciola hepatica*. High-resolution NMR spectra were obtained in 6 hours under anaerobic conditions (Mansour *et al.*, 1982). This is ample time to study the effects of different drugs on the intermediate metabolism of these parasites. One of the advantages of NMR is that the drugs are tested on the same live parasites that have been used during a control period of time. Changes in properties such as pH and ion concentrations can be monitored in the intracellular milieu. New and unusual metabolic intermediates can be found. An example is the identification of L-α-glycerophosphoryl choline (GPC) as a major metabolic intermediate in *Fasciola hepatica* (Mansour *et al.*, 1982). GPC was subsequently identified in *Ascaris suum* and appears to be associated with the effect of acetylcholine and its agonists, carbachol and levamisole, on the muscle contraction in *Ascaris* (Arevalo & Saz, 1992). ^{31}P-NMR can also be used to identify and quantify free cytosolic phosphate metabolites. Intracellular pH can be measured from the chemical shift of inorganic phosphate. Under resting conditions the pH in the parasites is maintained at 6.87 ± 0.05. Significant intracellular acidosis (to pH 6.2) develops with stimulation of motility by serotonin (5-hydroxytryptamine) (Matthews & Mansour, 1987).

^{31}P-NMR was also used to study the effects of mebendazole and dinitrophenol on the tapeworm *Hymenolepis diminuta* (Thompson, Platzer, & Lee, 1987). The effects of perfusing each of the drugs for 2.5 hours were determined, during which time the worms were viable as measured by the ^{31}P-NMR spectrum of the level of ATP. Whereas mebendazole had little effect on the nucleotide triphosphate level, 2,4 dinitrophenol (an oxidative phosphorylation uncoupler) caused a rapid decline in nucleotide triphosphate level. The conclusion drawn was that mebendazole does not exert its primary effect on oxidative phosphorylation.

The nature of different reactions involved in a metabolic pathway can be studied using [1-^{13}C glucose]. In *Fasciola hepatica* it was determined that acetate, a glycolytic end product in this parasite, is synthesized from pyruvate generated by the malic enzyme reaction rather than directly by catalysis of phosphoenol pyruvate by pyruvic kinase (Matthews *et al.*, 1985, 1986).

^{1}H-NMR is preferred over ^{31}P or ^{13}C to study glucose metabolism in parasites. Glucose and its metabolites have high natural abundances of protons. ^{1}H-NMR has a greater sensitivity and a higher speed of data acquisition compared to ^{31}P or ^{13}C methods. ^{1}H-NMR has been used to study the glucose metabolism of *Giardia lamblia*, a protozoal parasite of the intestinal tract of humans (Edwards *et al.*, 1989). The NMR spectra showed that axenically cultured trophozoites utilized glucose to produce acetate, ethanol, and alanine. The first two products were not a surprise; however, the production of sizable amounts of alanine had not previously been reported (see Chapter 6).

The examples given above illustrate aspects of the unique potential of NMR that should complement other molecular and biochemical studies on antiparasitic agents. NMR is dynamic and noninvasive and allows serial measurements to be made on a single sample. It can produce information about the intracellular microenvironment such as pH and illuminate the interaction of macromolecules with the parasite. It is an additional tool for discovering new metabolic pathways and their regulation in parasites. It should play an increasing role in the study of the effects of chemical agents on different metabolic pathways.

Use of Recombinant DNA Technology for Studies on Drug Targets

Since the early 1980s there have been great advances in the biology of parasitic organisms through the techniques of molecular biology and genetics. Genes that encode for most potential protein targets can be cloned in bacterial or eukaryotic cells. There are now available cDNA libraries for major parasitic protozoa and helminths, which can be screened with either heterologous probes or oligonucleotide probes prepared on the basis of amino acid sequences of the protein. cDNA libraries from helminths can be enriched by starting with material from a specific tissue of the parasite. This was shown in the case of the genital complex dissected from *F. hepatica* (Zurita *et al.*, 1987) or from the pharynx and nerve cord tissues of adult *A. suum* (Geary *et al.*, 1993). The nucleotide sequence of the isolated clone can provide the 5' and 3'-untranslated regions as well as the open reading frame. The deduced amino acid sequence is obtained from the open reading frame of the nucleotide sequence. It is now a routine practice to compare the sequence obtained with the other cloned genes. The next step is to express the gene in *E. coli* or other suitable cells (insect cells) using an appropriate vector. The expressed protein can be purified and the protein can be further characterized biochemically. An antiserum prepared against the protein provides the investigator with an immunological tool to gain more information about the expressed protein.

Tissue localization of the gene and the relevant protein adds more information about their physiological function. This is usually performed using a ^{32}P-random primed probe of the cDNA clone or ^{32}P-end-labeled synthetic oligonucleotide that will hybridize with the mRNA of the isolated clone. Hybridization is carried out on a cross section of the parasitic helminth or on intact cells in the case of protozoa. Cytological localization of the gene can be done using phase-contrast microscopy or a bright field with different degrees of magnification. Polyclonal antibodies labeled by fluorescent probes can also be used to identify antigen presence and its localization in the parasite. Because of the comparatively large size of parasitic helminths and the complexity of their internal organs, subcellular localization of genes is especially important to ascertain function. This was demonstrated in the case of an eggshell protein gene of *F. hepatica*, which was shown to be expressed not only in the eggshell but also in developing and mature vitelline cells (Zurita, Bieber, & Mansour, 1989).

Identification of Potential Drug Targets and Candidate Antigens for Vaccines Using Expressed Sequence Tags

The chromosomes of many parasites are too large to be separated by pulse-field electrophoresis. It is, therefore, not yet possible to prepare chromosome-specific libraries. For this reason a new strategy is based on identifying as many new genes as possible for future placement on chromosome maps. DNA of random clones obtained from cDNA libraries representing all stages of development are being sequenced to provide 100–400 nucleotides (usually from the 5' end). These random genes are identified as *expressed sequence tags* (ESTs). Each gene is represented in a cDNA library in proportion to the abundance of its mRNA. Therefore, genes encoding housekeeping enzymes are highly expressed. In contrast, genes involved in differentiation or development are rare. In addition to the use of cDNA libraries, genomic libraries with large fragment clones can be constructed in BAC (bacterial artificial chromosome) or YAC (yeast artificial chromosome). These are used for chromosome mapping through localization of large fragment clones that may contain known gene markers. The use of these different techniques has allowed extensive gene discovery at low cost (Williams & Johnston, 1999).

The World Health Organization (WHO) has established operational procedures for generating and analyzing EST data sets. To maintain a high gene discovery rate from EST information it is important to assess redundancy of sequences for each library. If redundancy exceeds one new gene per three ESTs generated the utility of the library should be reassessed. There is always an excess of certain genes such as those for contractile muscle or egg formation. Those could be screened out by hybridization or by using libraries from other stages. ESTs that are derived from the same gene are grouped together into clusters and the consensus sequence is deduced. Clustered ESTs are then subjected to bioinformatics processing using BLAST or other algorithms to search out motifs and patterns. Information from these studies is usually integrated with reported biological studies on the parasite to identify prime candidate genes for further testing.

One of the difficulties in carrying out genomic research on parasites has been the lack of genetic maps. It is almost impossible to expect complete genome sequences for all known helminth parasites. WHO is encouraging the sequencing of the *Schistosoma mansoni* and *Brugia malayi* genomes as models.

Information is usually compared with the sequences from the completed model organism, *C. elegans*, or with sequences from other pathogenic helminths. It has been reported that 40% of the genes identified in *Brugia* are novel genes that have no detectable identification with known sequences. Deducing the functions of other identified EST clustered genes is not possible from comparison with the *C. elegans* data set. These unidentified genes may very well include some new targets or could encode for unique antigenic components. These are the genes that deserve further studies of their structures and functions (Blaxter *et al.*, 1999). The WHO project already has a wealth of data and resources available online, including the parasite-genome computing resource http://www.ebi.ac.uk/parasites/genomecompute.html.

Time and effort can be saved by using the genomic information from organisms that have been systematically studied to learn about genes of parasites from related taxonomic groups without building a complete sequence map as described above. Detailed information gathered on a gene that was found to code for a new drug target in one species can be used to search for a corresponding gene in other species. It has been suggested that it would be advisable to obtain significant EST data sets of one or two species from each phylum. Even data from sets of 120–220 ESTs of representative parasitic nematodes could give us some idea of the proportion of novel genes (Blaxter *et al.*, 1999).

Genomic Analysis of Parasitic Helminths

Several research programs for studying parasite genomes and international arrangements for making the information available worldwide have been established. One reason for the great excitement about these programs is the understanding that genomic information about parasites of medical importance will play an important role in the discovery of new drug targets and candidate antigens for vaccines. A special symposium arranged by the British Society for Parasitology was held in September, 1998. Reviews of this symposium were published in a supplement to *Parasitology*, Volume 118 (1999). Studies on parasite genomes lag far behind those of mammals, viruses, bacteria, and the model free-living nematode *C. elegans*. Having the complete genome sequence contributes a great deal to understanding the functioning of different genes and the proteins they encode. Much of the sequence information of parasite genes was gained by the cloning of genes coding for proteins that were thought to be good candidate antigens for vaccines. Genes of several key enzymes that had been found to play an important role in the metabolism of certain parasites have also been cloned. *S. mansoni* has a complex genome of 270 million base pairs with seven pairs of autosomes and one pair of heteromorphic sex chromosomes. The *Brugia* genome has 100 million base pairs (bp) on five pairs of chromosomes with XY sex determination. This size is comparable to that of *C. elegans*.

Caenorhabditis elegans Genome

C. elegans was chosen initially as a model organism for studying the nervous system in nematodes. In addition to serving that purpose *C. elegans* is now used as a model organism for parasitologists interested in parasitic nematodes. *C. elegans* is a small free-living bacteriovorous nematode that is easy and inexpensive to keep in the laboratory. It is responsive to many of the antinematodal agents used against parasitic nematodes. Research at many laboratories has uncovered intricate details of anatomical and neurobiological development at the single cell level. In addition, several thousand loci have been identified by mutational genetics. The complete genome consists of a sequence of 100 million bp, arranged in five autosomes, and one sex chromosome (Blaxter *et al.*, 1999).

Genome Analysis of Trypanosomatids

Techniques similar to those described above are currently being employed to analyze genomes from three different species of trypanosomatids, *T. brucei*, *T. cruzi*, and *Leishmania major* (Blackwell & Melville, 1999). The TREU 927/4 strain of *T. brucei* has the smallest (53.4 megabase (Mb)) genome and has been chosen for genome analysis and sequencing (Melville, 1997, 1998). The strategy used in this and many other genome projects includes analysis of the chromosomal karyotypes metaphase by pulse-field electrophoresis. Physical mapping using large DNA libraries (cosmid, pacmid P_1, BAC, and YAC) is also being done. Partial cDNA sequence analysis is used to develop sets of ESTs for gene discovery and for use as markers in physical mapping and genomic sequencing. The *T. cruzi* genome has an estimated diploid gene size of ~87 Mb. On chromosome 3 of *T. cruzi* the first 93.4 kilobase (kb) of sequence was shown to contain 20–30 novel genes as well as several repeat elements including a novel chromosome

3-specific 40-bp repeat. Here the genes appear to be organized into two long clusters containing multiple genes on the same strand (Andersson *et al.*, 1998). A potential technique for monitoring transcription products from open reading frames employs microarrays of cDNAs on chips. This approach has been taken by Lashkari *et al.* (1997), using yeast microarrays for genome-wide parallel and gene expression analysis. Proteomics has also been used to analyze expression of different genes and to identify and characterize the corresponding proteins and their levels in *Leishmania* (Ivens & Blackwell, 1999).

Genome Analysis of *Plasmodium falciparum*

Because malaria is becoming drug resistant and the mosquito vector is becoming resistant to insecticides, more people than ever are becoming infected. There is an urgent need for the discovery of new targets for chemotherapeutic agents and candidate antigens for research and development of vaccines. Of the four species belonging to the genus *Plasmodium* that cause malaria in humans, infection with *P. falciparum* has the highest incidence of morbidity and mortality. For these reasons *P. falciparum* was chosen as the first model for genomic analysis. The genetic sequencing, which began in May 1996, is expected to be complete by 2002. Up-to-date sequence data as well as partially assembled contigs (contiguous sequences of DNA) are available at http://plasmodiumdb.cis.upenn.edu. The *P. falciparum* genome is 30 Mb in size. It contains fourteen chromosomes ranging in size form 0.65 to 3.4 Mb. Initially chromosome 2 (one of the smallest) of *P. falciparum* was chosen for genomic analysis (Gardner *et al.*, 1998). Sequencing was done using the "shotgun" strategy previously used for sequencing of *Haemophilus influenzae* (Fleischmann *et al.*, 1995). Computational methods were developed to assemble hundreds of thousands of 300–500 bp cDNA sequences. This provided a sufficient number of fragments that can be sequenced and assembled to produce a complete chromosome sequence. The *P. falciparum* chromosome 2 is 947 kb in length and contains ninety predicted proteins with no detectable homologs. A family of genes that encode for surface proteins, dubbed rifins, has been identified. Although the function of these encoded proteins has not yet been verified, their predicted structure indicates that they possess a transmembrane moiety as well as an extracellular domain that could play a role in the antigenic variation process in the parasite (Gardner *et al.*, 1998). These proteins could be of great interest as antigen candidates for vaccines or drug targets in *P. falciparum*.

Genome studies of *P. falciparum* are still continuing. Each of the parasite chromosomes has been assigned to a participant in a consortium supported by the United Kingdom Wellcome Trust, United States NIH, or other institutions (Hoffman *et al.*, 1997). The basic philosophy guiding the group is that genome research is the most cost-effective means of discovering new targets for chemotherapeutic agents and proteins with likely vaccine possibilities. It will not be too long before the scientific community is armed with the genomic sequences of some of the most dreadful parasites humankind has ever known.

While this book was being written several reviews have appeared discussing the great expectations that scientists all over the world have for the tremendous advances that will be hastened by completion of the malaria genome (Carucci, 2000; Horrocks *et al.*, 2000; Macreadie *et al.*, 2000). These reviews give an encouraging picture of the future of malaria research. For scientists from the developed world where the laboratory resources and financial support are available, the expectations are that there will be systematic investigations of development, differentiation, and energy metabolism of the parasite and response of the host immune system. The interactions of the parasite with the host will be examined using comparative genomics. It will be possible to study the complex interactions of hundreds of genes and proteins that are essential for maintaining the parasitic life rather than having to focus on one gene or enzyme at a time. Certainly studies based on the malaria genome should enhance our understanding of the biology of parasites and clarify the host–parasite relationship. For scientists in the developing countries where resources are very limited, this information will still be valuable. These scientists have first-hand knowledge of the problems associated with this dreadful disease and have immediate incentives to use any new information productively. DNA sequences generated by the Malaria Genome Project as well as the working draft of the human genome should facilitate the identification of novel drug targets. In addition, better understanding of the molecular mechanisms of action of currently used drugs and the mechanisms of drug resistance should enhance our ability to find more selective, efficient, and affordable chemotherapeutic agents.

Structural Genomics

The prodigious amount of information becoming available about the DNA sequences of both parasitic organisms and their hosts has flooded current databases at a startling velocity. However, the analysis of these data does not yet give enough information about the structure of the proteins encoded or about their functions. Structural genomics is an upcoming field that tackles the problems of assigning structure and functions to the deduced proteins. In spite of the great advances in crystallographic techniques and NMR spectroscopy in identifying

3-D structure, only about 1% of the estimated proteins in humans has been analyzed. A large number of genome sequences share no sequence identity to those proteins for which the 3-D structure is known. Many of these have unknown functions. In many cases, finding the function of the unidentified protein is essential for understanding details of an important metabolic pathway and its connection to survival of the organism. An approach that has been reported by Sung-han Kim and his group at Berkeley has been successful in identifying one of these proteins from a hyperthermophile, *Methanococcus jannaschii* (Zarembinski *et al.*, 1998). These investigators first determined the 3-D structure by X-ray crystallography. The 3-D structure of the unknown protein (not the amino acid sequence) when compared to other proteins in the structure database (Protein Data Bank) showed that the protein had protein-fold similarities to an ATP-binding molecular switch for a cellular process. This protein switch controls hydrolysis of ATP in the presence of another component in the bacteria. The nature of this protein was verified subsequently by biochemical means. This is a good example of the structural genomics approach to identifying an unknown function of a protein.

The new science of structural genomics is now being developed to find the relationship between the sequence of a gene and the structure and function of the protein for which it codes. This structural information entails cataloging the different ways that proteins fold. It is now estimated that of 1000 protein folds, only half are known. It is optimistically anticipated that the structural gap will be filled soon, at least for soluble proteins. For a review see Rost (1998). Computer programs are being developed that will be able to predict the 3-D structure of a protein from a simple linear sequence of DNA. Identifying the 3-D structure of a large number of proteins should advance the technology for drug design and discovery (Service, 2000).

Proteomics

Whereas the field of genomics deals essentially with DNA sequencing, proteomics is the study of the nature of different proteins encoded by the genes. Results obtained by genomics about cellular genes are static; proteomics deals with the more dynamic processes carried out by proteins. (See the review of Fields (2001).) Proteomics includes identification and quantitation of proteins and their cellular locality and modifications of their structure and function. *Proteome* is a fairly new word in the world of molecular biology and is defined as the total protein complement of a genome. It has been extended to include the proteins synthesized in a cell type, tissue, or developmental stage (Ashton,

Curwen, & Wilson, 2001). The old understanding that one gene codes for one protein is no longer valid. It is now recognized that one gene can encode for different proteins by the process of alternative splicing of the mRNA, by varying translation start or stop sites, or by shifting the reading frame of the 3′ side of the splice. This would give rise to proteins with sequences that are different from those encoded by the DNA. Proteins can be covalently modified (by, e.g., phosphorylation), which could result in drastic changes in their biochemical function in the cell. Such modification could result in significant change in the kinetics or even cause inactivation of the enzyme. Post-translational modification can also include linking the enzyme to glycosylphosphatidylinositol to make the protein change location to integrate into cell membranes. Other ways of changing the nature of the proteins include glycosylation, acetylation, ubiquitination, and farnesylation. These post-translational changes to the proteins can very well influence the process of finding new drug targets.

New knowledge of proteomics has made us recognize that the level of a protein does not always reflect mRNA level (Fields, 2001). Furthermore, identification of an open reading frame does not necessarily guarantee the existence of the protein. Another confusing possibility that researchers encounter is that a single protein may be involved in more than one process and, conversely, similar functions may be carried out by multiple proteins.

Essential technologies needed for proteomics include protein expression patterns from cells or tissues by two-dimensional electrophoresis (2-DE) or other techniques that characterize large sets of proteins. In the first direction of the electrophoresis proteins are usually separated by electrical charge and in the second direction they are separated by molecular size. The 2-DE technique may not be sensitive enough to identify all of the potential 10,000 protein spots. Other procedures that characterize larger sets of proteins can be used. A technique that allows rapid identification of small amounts of large numbers of proteins, such as ultrasensitive mass spectrometry coupled to throughput functional screening assays, is necessary. The availability of innovative computational equipment is also necessary to process, analyze, and interpret the large amounts of data. There are software packages and other bioinformatic databases for identification of proteins. These could include 3-D structure, potential post-translational modification, and physico-chemical characterization.

There are a few clues that are being used to identify new proteins in a cell (Fields, 2001). One of these is that proteins prefer to be near other proteins in the cell. Efficient algorithms are now available to assign functions to previously known proteins that do not depend on amino acid similarity. The cellular sites of the unknown proteins could also give clues to their identities. A protein found to be interacting with actin or with glycogen should suggest its location and give some idea about its functions. More biochemical or biophysical experimentation should be employed to verify the localization and functions of the protein (WHO, 2000).

More knowledge about the proteomes of parasites at different stages of their life cycle should provide fruitful leads to multiple, rather than only single, proteins involved in growth, development, and virulence. Some of these proteins could be used as prospective antiparasitic targets or candidate antigens for vaccines.

Identification of proteins that are unique to the parasites using proteomics could be useful for diagnosis of parasitic infections. Some parasites, such as *Schistosoma mansoni* and *S. hematobium* are known to shed their outer tegumental membranes during development in the host (see Chapter 8). Identification of these proteins in the blood or urine (in the case of *S. hematobium*) could lead to early diagnosis of infection. This might prevent subsequent squamous cell carcinoma of the bladder, which is a not uncommon consequence of infection with *S. hematobium* (Banks *et al.*, 2000).

Billions of humans worldwide who are infected with protozoa or helminths are currently being treated with chemotheraeutic agents of high toxicity and low efficacy. Of the few antiparasitic agents that are currently in our therapeutic armamentarium none have been discovered using rational drug design. During the past decade there has been an upsurge of new technology for identifying new selective targets that are either unique to the parasite or are quite different from their homologs in the host. In addition, there has been a radical change in the technology for selectively identifying lead compounds from thousands of substances produced by combinatorial chemistry. A great incentive for using rational approaches is that, although the initial costs may be considerable, they may be less in the long run. Much of this new technology, already in use for finding medicines for other diseases, can be adapted in the search for better chemotherapies against parasites.

REFERENCES

Andersson, B., Aslund, L., Tammi, M., Tran, A. N., Hoheisel, J. D. & Pettersson, U. (1998). Complete sequence of a 93.4-kb contig from chromosome 3 of *Trypanosoma cruzi* containing a strand-switch region. *Genome Res, 8*(8), 809–816.

Arevalo, J. I. & Saz, H. J. (1992). Effects of cholinergic agents on the metabolism of choline in muscle from *Ascaris suum. J Parasitol, 78*(3), 387–392.

Ashton, P. D., Curwen, R. S. & Wilson, R. A. (2001). Linking proteome and genome: How to identify parasite proteins. *Trends Parasitol, 17*(4), 198–202.

Banks, R. E., Dunn, M. J., Hochstrasser, D. F., Sanchez, J. C., Blackstock, W., Pappin, D. J. & Selby, P. J. (2000). Proteomics: New perspectives, new biomedical opportunities. *Lancet, 356*(9243), 1749–1756.

Blackwell, J. M. & Melville, S. E. (1999). Status of protozoan genome analysis: Trypanosomatids. *Parasitology, 118*(Suppl), S11–14.

Blaxter, M., Aslett, M., Guiliano, D. & Daub, J. (1999). Parasitic helminth genomics. Filarial Genome Project. *Parasitology, 118*(Suppl), S39–51.

Carucci, D. J. (2000). Malaria research in the post-genomic era. *Parasitol Today, 16*(10), 434–438.

Cohen, F. E., Gregoret, L. M., Amiri, P., Aldape, K., Railey, J. & McKerrow, J. H. (1991). Arresting tissue invasion of a parasite by protease inhibitors chosen with the aid of computer modeling. *Biochemistry, 30*(47), 11221–11229.

Edwards, M. R., Gilroy, F. V., Jimenez, B. M. & O'Sullivan, W. J. (1989). Alanine is a major end product of metabolism by *Giardia lamblia*: A proton nuclear magnetic resonance study. *Mol Biochem Parasitol, 37*(1), 19–26.

Fairlamb, A. H. & Cerami, A. (1985). Identification of a novel, thiol-containing co-factor essential for glutathione reductase enzyme activity in trypanosomatids. *Mol Biochem Parasitol, 14*(2), 187–198.

Fields, S. (2001). Proteomics. Proteomics in genomeland. *Science, 291*(5507), 1221–1224.

Fleischmann, R. D., Adams, M. D., White, O., Clayton, R. A., Kirkness, E. F., Kerlavage, A. R., Bult, C. J., Tomb, J. F., Dougherty, B. A., Merrick, J. M. *et al.* (1995). Whole-genome random sequencing and assembly of *Haemophilus influenzae* Rd. *Science, 269*(5223), 496–512. [See comments.]

Gardner, M. J., Tettelin, H., Carucci, D. J., Cummings, L. M., Aravind, L., Koonin, E. V., Shallom, S., Mason, T., Yu, K., Fujii, C., Pederson, J., Shen, K., Jing, J., Aston, C., Lai, Z., Schwartz, D. C., Pertea, M., Salzberg, S., Zhou, L., Sutton, G. G., Clayton, R., White, O., Smith, H. O., Fraser, C. M., Hoffman, S. L. *et al.* (1998). Chromosome 2 sequence of the human malaria parasite *Plasmodium falciparum*. *Science, 282*(5391), 1126–1132.

Geary, T. G., Winterrowd, C. A., Alexander-Bowman, S. J., Favreau, M. A., Nulf, S. C. & Klein, R. D. (1993). *Ascaris suum*: Cloning of a cDNA encoding phosphoenolpyruvate carboxykinase. *Exp Parasitol, 77*(2), 155–161.

Hoffman, S. L., Bancroft, W. H., Gottlieb, M., James, S. L., Burroughs, E. C., Stephenson, J. R. & Morgan, M. J. (1997). Funding for malaria genome sequencing [letter; comment]. *Nature, 387*(6634), 647.

Horrocks, P., Bowman, S., Kyes, S., Waters, A. P. & Craig, A. (2000). Entering the post-genomic era of malaria research. *Bull World Health Organ, 78*(12), 1424–1437.

Hunter, W. N. (1997). A structure-based approach to drug discovery; crystallography and implications for the development of antiparasite drugs. *Parasitology, 114*(Suppl), S17–29.

Ivens, A. & Blackwell, J. (1999). The *Leishmania* genome comes of age. *Parasitol Today, 15*, 225–221.

Lashkari, D. A., DeRisi, J. L., McCusker, J. H., Namath, A. F., Gentile, C., Hwang, S. Y., Brown, P. O. & Davis, R. W. (1997). Yeast microarrays for genome wide parallel genetic and gene expression analysis. *Proc Natl Acad Sci USA, 94*(24), 13057–13062.

Li, C. M., Tyler, P. C., Furneaux, R. H., Kicska, G., Xu, Y., Grubmeyer, C., Girvin, M. E. & Schramm, V. L. (1999). Transition-state analogs as inhibitors of human and malarial hypoxanthine-guanine phosphoribosyltransferases. *Nat Struct Biol, 6*(6), 582–587. [See comments.]

Macreadie, I., Ginsburg, H., Sirawaraporn, W. & Tilley, L. (2000). Antimalarial drug development and new targets. *Parasitol Today, 16*(10), 438–444.

Mansour, T. E., Morris, P. G., Feeney, J. & Roberts, G. C. (1982). A ^{31}P-NMR study of the intact liver fluke *Fasciola hepatica*. *Biochim Biophys Acta, 721*(4), 336–340.

Matthews, P. & Mansour, T. (Eds.). (1987). *Applications of NMR to Investigation of Metabolism and Pharmacology of Parasites*. In A. J. MacInnis (Ed.) *Molecular Paradigms for Eradicating Helminthic Parasites*. UCLA Symposium on Molecular and Cellular Biology, New Series, Vol. 60(pp. 477–492) New York: Alan R. Liss.

Matthews, P. M., Shen, L. F., Foxall, D. & Mansour, T. E. (1985). ^{31}P-NMR studies of metabolite compartmentation in *Fasciola hepatica*. *Biochim Biophys Acta, 845*(2), 178–188.

Matthews, P. M., Foxall, D., Shen, L. & Mansour, T. E. (1986). Nuclear magnetic resonance studies of carbohydrate metabolism and substrate cycling in *Fasciola hepatica*. *Mol Pharmacol, 29*(1), 65–73.

Melville, S. (1998). The African trypanosome genome project. Focus on the future. *Parasitol Today, 14*, 128–130.

Melville, S. E. (1997). Parasite genome analysis. Genome research in *Trypanosoma brucei*: Chromosome size polymorphism and its relevance to genome mapping and analysis. *Trans R Soc Trop Med Hyg, 91*(2), 116–120.

Ring, C. S., Sun, E., McKerrow, J. H., Lee, G. K., Rosenthal, P. J., Kuntz, I. D. & Cohen, F. E. (1993). Structure-based inhibitor design by using protein models for the development of antiparasitic agents. *Proc Natl Acad Sci USA, 90*(8), 3583–3587.

Rost, B. (1998). Marrying structure and genomics. *Structure, 6*(3), 259–263.

Service, R. F. (2000). Structural genomics offers high-speed look at proteins. *Science, 287*(5460), 1954–1956.

Thompson, S. N., Platzer, E. G. & Lee, R. W. (1987). In vivo ^{31}P-NMR spectrum of *Hymenolepis diminuta* and its change on short-term exposure to mebendazole. *Mol Biochem Parasitol, 22*(1), 45–54.

WHO. (2000). Proteomics. *Nat Biotechnol, 18 Suppl*(3), IT45–46.

Williams, S. A. & Johnston, D. A. (1999). Helminth genome analysis: The current status of the filarial and schistosome genome projects. Filarial Genome Project. Schistosome Genome Project. *Parasitology, 118*(Suppl), S19–38.

Zarembinski, T. I., Hung, L. W., Mueller-Dieckmann, H. J., Kim, K. K., Yokota, H., Kim, R. & Kim, S. H. (1998). Structure-based assignment of the biochemical function of a hypothetical protein: A test case of structural genomics. *Proc Natl Acad Sci USA, 95*(26), 15189–15193.

Zurita, M., Bieber, D. & Mansour, T. E. (1989). Identification, expression and in situ hybridization of an eggshell protein gene from *Fasciola hepatica*. *Mol Biochem Parasitol, 37*(1), 11–17.

Zurita, M., Bieber, D., Ringold, G. & Mansour, T. E. (1987). Cloning and characterization of a female genital complex cDNA from the liver fluke *Fasciola hepatica*. *Proc Natl Acad Sci USA, 84*(8), 2340–2344.

3

Energy Metabolism in Parasitic Helminths

Targets for Antiparasitic Agents

During the first half of the twentieth century studies on the biochemistry of energy production showed that there are great similarities among diverse species. The discoveries of complex pathways and the details of the enzymes involved encouraged biochemists to investigate the metabolism of many different parasites. Pharmacologists who were interested in the chemotherapy of these parasites looked at these studies as a way of finding reactions that are unique to the parasite and that the enzymes catalyzing these reactions could be targets for new antiparasitic agents.

Glucose and/or glycogen constitute the main sources of energy to produce ATP for parasitic helminths. When these parasites live in a milieu that is rich with sugars their metabolism is adapted to utilizing large amounts of glucose. They resort to the inefficient degradation of glucose through anaerobic metabolism, which produces only low levels of ATP plus excreted metabolic products such as lactic acid and/or fatty acids. The utilization of large amounts of glucose compensates for the inefficient nature of the parasites' anaerobic metabolism. Many antiparasitic agents owe their successful antiparasitic action to a selective inhibition of energy production by the parasite. A good understanding of the nature of different carbohydrate metabolic reactions in parasitic helminths is a solid basis for choosing appropriate targets for new chemotherapeutic agents.

Life Cycles of Representative Helminths

Although this chapter focuses on the adult stages of the parasites discussed, a brief description of the life cycles of three representative species is included. This is done for the sake of clarifying the developmental background and how it may influence the discovery of new antiparasitic agents. More detailed information on the life cycles of these and other parasites can be obtained from a good parasitology text (e.g., Smyth, 1994).

Schistosomes (Blood Flukes)

The blood flukes are flatworms, a large family of parasitic helminths that belong to the class *Trematoda*. The main species of blood flukes that infect humans are *Schistosoma mansoni*, *S. japonicum*, and *S. hematobium*. These are diecious organisms (having both males and females) and inhabit the mesenteric-portal venules (*S. mansoni* and *S. japonicum*) or the venous plexus surrounding the urinary bladder (*S. hematobium*). The mature adults are ~13 mm long, are continuously coupled during their adult life span in the venules, and mate frequently. The female worms lay eggs that are fully embryonated. Mature eggs penetrate into the lumen of the intestine or the urinary bladder, depending on the species. Once the eggs leave the host with the feces or the urine they hatch in stagnant water to become miracidia, which seek out the appropriate snail hosts. Inside the snail the miracidia develop to sporocysts and eventually to cercariae, which leave the snail to swim in the water. They eventually enter the mammalian host by penetrating the skin. Once in the host they are referred to as schistosomula and are carried by the bloodstream to the pulmonary capillaries of the lung and then through the heart to finally lodge in the portal venules or in the case of *S. hematobium* in venules around the urinary bladder.

The Liver Fluke *Fasciola hepatica*

The liver fluke *Fasciola hepatica* is an oval hermaphroditic flatworm about 35-mm long and 15 mm wide at the center. The adult parasites live in the bile ducts of cattle, sheep, or humans. The eggs pass from the bile duct to the intestine and are excreted from the host's body with the feces. In stagnant water the miracidia emerge from the eggs and infect a snail host. After two months the metacercariae leave the snail and become encysted and attached to water plants. Once the metacercariae are ingested by the final host they hatch in its intestine. From there they burrow through the gut wall and into the peritoneum, then into the liver, and eventually into the bile ducts.

The Intestinal Nematode *Ascaris suum*

Ascaris suum is a diecious roundworm about 15–30 cm in length and about 0.5 cm in diameter at the midsection. *Ascaris lumbricoides* is the human parasitic roundworm and appears to have only minor morphological differences from the pig *Ascaris*. Infection occurs through ingestion of eggs, which hatch in the host's intestine. From there the larvae burrow through the intestinal wall and enter the circulation going through the heart to the lungs. They work their way up the trachea to the pharynx and the mouth, where they are swallowed by the host and reenter the digestive system. The larvae complete their development to adults in the host's small intestine.

Energy Metabolism of *S. mansoni*

Glucose taken up by the parasite from the host or intracellular glycogen stores serve as the main sources of energy in schistosomes. The miracidia and cercariae are predominately aerobic (Tielens *et al.*, 1992). Consequently, CO_2 is the main metabolic product whereas lactic acid accounts for only 10–30% of the metabolized carbohydrates. Transformation of miracidia to sporocysts in the snail is accompanied by a greater percentage of glucose metabolized to lactic acid. As the schistosomula mature to adult worms in the host, energy production shifts to anaerobic metabolism. Glucose metabolized through the glycolytic pathway to lactic acid is the main source of energy for these parasites (Bueding, 1950). The amount of glucose utilized in 1 hour is as high as 20% of the parasite's dry weight (approximately 10% of wet weight). Lactic acid can account for almost all of the utilized glucose. According to this early research, the rates of glucose utilization and lactic acid production are the same whether the organisms are incubated under aerobic or anaerobic conditions. Accordingly, there is no Pasteur effect (inhibition of glycolysis by respiration) (Pasteur, 1861) that can be detected. The parasites use oxygen but its role in the parasite's energy metabolism has been disputed. Levels of ATP in the parasites are the same under aerobic or anaerobic conditions (Bueding & Fisher, 1982). Since these early experiments by Bueding, it has been generally accepted that these parasites are primarily homolactate fermenters and that the oxygen utilized has a function other than that of energy production from glucose. Evidence for the participation of aerobic glucose metabolism in schistosomes was reported by Coles (1973), who showed that lactic acid produced by the worms under anaerobic conditions was increased when compared with schistosomes kept under air. Coles concluded that at least a quarter of the parasites' energy comes from oxidative metabolism (Coles, 1973). It has also been reported that the worms have a coupled functional cytochrome system and functional citric acid cycle. These systems may not be used in the adult schistosomes for oxidative phosphorylation. Since these reports, the contribution of aerobic metabolism for energy in schistosomes has been argued in the literature. More recently, van Oordt *et al.* (1985) reported that under aerobic incubation conditions using D-[6-^{14}C] glucose a considerable amount of $^{14}CO_2$ was produced, indicating that the Krebs cycle was operative. In contrast, under anaerobic conditions or in the presence of cyanide no $^{14}CO_2$ production occurred, indicating that the carbon dioxide production is coupled to oxidative phosphorylation. The contribution of aerobic metabolism in the parasites could be related to egg production by the females. When cultured under nitrogen parasites do not lay eggs and thus fail to show an increase in the glycolysis that would normally be used for egg production. More investigations are needed to verify this explanation (Coles, 1973). Because glycolysis plays a crucial role for energy production in schistosomes considerable attention has been devoted to the various components of the pathway.

Glucose Transporters

The first phase in the regulation of glycolysis is the control of glucose uptake from the outside environment. The process of transporting glucose across the cell plasma membrane is facilitated by several glucose transporters. Sugar transporters are ubiquitous among different animal species. Five genes have been identified in mammals that encode for transporters, GLUT 1 to 5, in different cells (Silverman, 1991). Each transporter consists of a single polypeptide chain, about 500 amino acids long. There is good homology among the structures of these proteins. Glucose binding to its site in the transporter induces conformational change that reorients the carrier in the membrane, which results in the translocation of the sugar to the inside of the cell.

Two full-length cDNAs that encode for glucose transporters in *S. mansoni* (SGTP$_1$ and SGTP$_4$) have been isolated and studied by Skelly and Shoemaker (Skelly *et al.*, 1994). Predicted amino acid sequences of the two proteins show a substantial homology with the sequences of mammalian, eukaryotic, and prokaryotic transporters. The schistosome sugar transporters are located in different parts of the tegumental membranes (the cytoplasmic structure that surrounds the entire worm). SGPT$_4$ is located in the outer tegumental membranes (Skelly & Shoemaker, 1996). Presumably, these are the transporters that transfer glucose from the host bloodstream into the tegumental syncytium (see Fig. 8.1 in Chapter 8). The other glucose transporters (SGTP$_1$) are concentrated primarily in the foldings of the basal membrane of the tegument and in the underlying musculature. Presumably, SGTP$_1$ transfers free glucose from the tegument into the interstitial fluids that are around the internal organelles of the parasite (Zhong *et al.*, 1995). The transporters, their location in the tegument, and their functions are discussed more fully in Chapter 8.

The Glycolytic Pathway

The next phase of glucose metabolism is its degradation via the different enzymatic reactions of the Embden–Meyerhof pathway (Fig. 3.1). The participating enzymes have been identified in many parasites and the reactions are the same as those found in mammals, yeast, and bacteria. Some of the rate-limiting enzymes from parasites have been purified and cloned and their roles as targets for antiparasitic agents have been characterized. Discussion of the nature of these key enzymes in the parasites and how they compare with those in the host are emphasized in this section.

Hexokinase. Hexokinase is the first enzyme in the glycolytic pathway. Mammalian hexokinases from muscle, liver, and other organs have been purified and their kinetic properties have been identified. There are four isozymes present in mammals, hexokinases I, II, III, and IV. The first three are single peptides with molecular masses of approximately 100 kDa. Hexokinase IV (better known as glucokinase) is expressed in the liver and has a molecular mass of $\sim$50 kDa. The kinetic properties of each isozyme play an important role in their regulation and may reflect specific physiological functions of different cells. Hexokinases I, II, and III have a higher affinity for glucose. Glucose 6-P, the product of the reaction, is important in the regulation of hexokinase and of glucose utilization. The sensitivity of different hexokinases to the inhibitory effect of glucose 6-P varies. Isoenzymes I and II are sensitive to inhibition by glucose 6-P whereas hexokinase III requires ten times as much as I and II. In contrast, glucokinase (type IV) is not sensitive to inhibition by glucose 6-P.

Hexokinase from *S. mansoni* has recently been purified, cloned, expressed in *E. coli*, and characterized (Shoemaker *et al.*, 1995). Its cDNA encodes for a 50-kDa protein. There is a strong homology between the binding sites for ATP and glucose in this enzyme and those reported for rat hexokinase. Southern hybridization indicates that the hexokinase gene is a single copy that is expressed in all the intermediate developmental stages of the parasite. The evidence indicates that this is the only hexokinase expressed in adult schistosomes. The enzyme has a high affinity to glucose. Glucose 6-P is a competitive inhibitor with respect to ATP, but it is less potent when compared to its inhibition of the mammalian heart enzyme. Schistosome hexokinase is unique since it is the first 50-kDa glucose 6-P–sensitive hexokinase (Tielens *et al.*, 1994). The schistosome enzyme

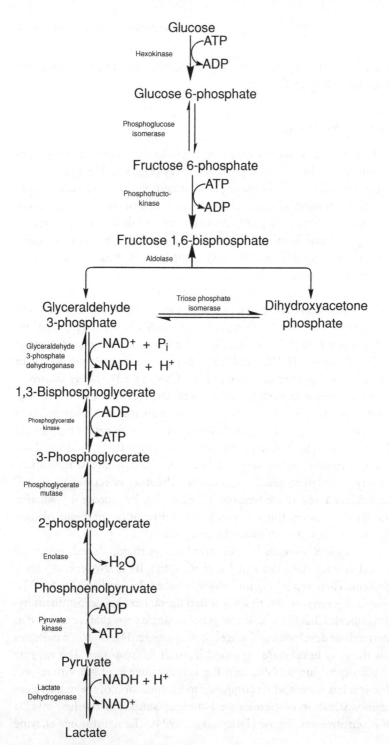

Figure 3.1. The Embden–Meyerhof scheme of glycolysis.

lacks a hydrophobic segment at the N-terminus and therefore does not bind to the mitochondria as do the mammalian hexokinase I and II isozymes. Because it is the only enzyme in the parasite available for glucose phosphorylation its selective inhibition could prove to be a good way of keeping the parasite from utilizing glucose from the host.

Inhibition of Glucose Utilization

5-Thio-D-glucose (5-TG) is an analogue of D-glucose that contains a sulfur atom in the ring instead of oxygen. The compound was found to be nontoxic to mammals (LD_{50} of 14 g/kg). As soon as it was synthesized evidence was presented that it has a diabetogenic effect when administered orally or intraperitoneally. The diabetogenic action appears to be due, in part, to the ability of 5-TG to interfere with cellular transport processes for D-glucose (Whistler & Lake, 1972). The effect was attributed to inhibition of the uptake of glucose by different tissues. Subsequently, 5-thioglucose was shown to have a rapid inhibitory effect *in vitro* on glucose metabolism in intact schistosomes, characterized by a reduction in motility (Tielens, Houweling, & Van den Bergh, 1985). The mechanism by which glucose enters the cells (presumably through $SGTP_4$) was inhibited by 5-TG. In addition, 5-TG was shown to be a competitive inhibitor of hexokinase in cell-free extracts of parasites. The inhibition was competitive with respect to glucose and could be the main contributory factor in inhibition of glucose metabolism. None of the other glycolytic enzymes or glycogen metabolism enzymes were affected by 5-TG. In spite of the fact that 5-TG was not toxic and that it selectively inhibited glucose transport and hexokinase, it did not appear to be a satisfactory schistosomicide. One reason is that its inhibitory effect on hexokinase is competitive with respect to glucose. It was pointed out that this inhibition may not be effective *in vivo* since the glucose concentration in the mesenteric venules where the adult schistosomes live is tenfold higher (5 mM) than the *in vitro* concentrations used for the experiments. The discovery of the inhibitory effect of 5-TG draws our attention to new targets (glucose transport and hexokinase) that deserve further studies using other glucose analogs.

5-THIO-D-GLUCOSE

Phosphofructokinase (PFK)

Regulation of PFK

At the outset of the studies on the regulation of glycolysis in the trematodes *S. mansoni* and *F. hepatica* experimental evidence accumulated that phosphofructokinase is a rate-limiting enzyme in the glycolytic pathways of both organisms. This was proven whether glycolysis was inhibited by the anthelmintic trivalent antimonials (Mansour & Bueding, 1954; Bueding & Mansour, 1957) or was activated by serotonin (Mansour & Mansour, 1962). In the 1960s mammalian PFK (because of its instability) was the only major glycolytic enzyme that had not been purified, which meant that its regulatory properties had not been studied. The great interest in knowing more about the PFK of parasites gave impetus to study mammalian PFK. This was important in defining differences between the host and parasite enzymes and the effect of antiparasitic agents.

Two regulatory mechanisms for PFK from different parasites have been discovered: regulation through allosteric kinetics of the enzyme and covalent phosphorylation of the enzyme by cyclic AMP-dependent protein kinase. Mammalian PFK, however, is regulated only through its allosteric kinetics (Mansour, 1972). The crucial ligand for regulation by allosteric kinetics is ATP. In addition to being a substrate, ATP also functions as an allosteric inhibitor by binding at a site different from the catalytic site (Setlow & Mansour, 1970). High levels of ATP result in an increase in the cooperative kinetics for fructose 6-P and inhibition of the enzyme. In mammals citrate plays an important role in aerobic metabolism by potentiating the inhibitory effect of ATP, thus increasing the cooperativity for fructose 6-P. However, neither *S. mansoni* nor *F. hepatica* PFKs are inhibited by citrate (Kamemoto *et al.*, 1987; Su, Mansour, & Mansour, 1996). This is consistent with the fact that both parasites have a predominantly anaerobic metabolism and the citric acid cycle is not involved significantly in their energy metabolism. AMP and fructose 2,6-bisP are the main allosteric activators of ATP-inhibited PFK in both mammals and parasites (Kamemoto *et al.*, 1987). Both activator ligands reduce the cooperativity of fructose 6-P and thus activate the enzyme. Activation of PFK by covalent phosphorylation with cAMP-dependent protein kinase was discovered in a few other parasites, including *F. hepatica* (Kamemoto & Mansour, 1986) and *A. suum* (Kulkarni *et al.*, 1987). Phosphorylated *F. hepatica* PFK is activated by a lower concentration of fructose 6-P than the nonphosphorylated enzyme. The phosphorylated activated enzyme can be further activated by the allosteric activators AMP and fructose 1,6-bisP (Kamemoto *et al.*, 1987). Mammalian PFK is phosphorylated by cyclic AMP-dependent protein kinase, but the phosphorylation does not increase enzyme activity.

```
                                 P
                                 |
Rabbit muscle:      Arg-Lys-Arg-Ser-Gly-Glu-Ala-Thr-Val-COOH


                                 P
                                 |
A. suum muscle:     Ala-Lys-Gly-Arg-Ser-Asp-Ser-Ile-Val-Pro-Thr


                            P
                            |
F. hepatica:        Arg-Ser-Thr-Met-Met-Iso-Pro-Gly-Met-Glu-Gly-Lys
```

Figure 3.2. Phosphorylation sites for phosphofructokinase from rabbit muscle (Kemp, 1969), *Ascaris suum* muscle (Kulkarni *et al.*, 1987), and *Fasciola hepatica* (Mahrenholz *et al.*, 1991).

The phosphorylation site in *F. hepatica* PFK was found to be a threonine, which is different from the usual serine (Fig. 3.2) (Mahrenholz, Hefta, & Mansour, 1991). The sequence of the phosphorylated peptide is not homologous to any previously determined PFK phosphopeptides. This is particularly surprising when it is compared to the phosphopeptides in which PFK is activated by phosphorylation. The *F. hepatica* phosphopeptide does fit the consensus sequence. However, the *Ascaris* phosphopeptide is particularly divergent when compared with the consensus sequence of other cAMP-dependent protein kinases (Kulkarni *et al.*, 1987). An explanation for the ambiguity regarding the *Ascaris* PFK should be resolved by obtaining more structural information about other related parasites to establish the degree of homology among these PFKs that are activated by phosphorylation. Such information may provide a structural explanation of why covalent phosphorylation stimulates enzyme activity in one case but not in another.

Activation of PFK and Glycolysis by Serotonin

Serotonin (5-hydroxytryptamine) is an indolalkylamine that is widely present among both mammals and invertebrates, including parasitic helminths. Its effect in stimulating the motility of trematodes (Mansour, 1957) and cestodes (Mettrick & Cho, 1981) is very striking (see Chapter 7). In addition, serotonin was shown to increase glucose uptake, glycogen breakdown, and lactic acid production. *Fasciola hepatica* was studied as a model organism for determining the mechanism of activation of glycolysis by serotonin. Originally, it was thought that the increase in glycolysis is a secondary response to the increase in parasite motility. This does not appear to be the case since addition of serotonin to cell-free homogenates of *F. hepatica* increased both glucose uptake and lactic acid production, though the effect was not as large as when using intact

SEROTONIN
(5-HYDROXYTRYPTAMINE)

R=H D-LYSERGIC ACID DIETHYLAMIDE (LSD)
R=Br 2-BROMO LSD (BOL)

organisms (Mansour & Mansour, 1962). This aspect of the research was done around the time when the late Earl Sutherland was gathering evidence that cyclic-3',5'-AMP acts as a second messenger for many mammalian hormones (e.g., epinephrine and glucagon) that control carbohydrate metabolism. Indeed, serotonin was shown to act in *F. hepatica* like epinephrine in mammals. Serotonin and not epinephrine was shown to activate adenylyl cyclase in *F. hepatica* (Mansour et al., 1960). Furthermore, serotonin and cyclic-3',5'-AMP increased glycolysis when added to cell-free homogenates (Mansour & Mansour, 1962). These studies on the effect of serotonin on *F. hepatica* served as the first non-mammalian example for Sutherland's conception of the unitarian function of cyclic-3',5'-AMP.

Serotonin and some of its analogs were the only bioamines investigated that increased the production of cAMP by adenylyl cyclase in *F. hepatica* (Northup & Mansour, 1978) and *S. mansoni* (Estey & Mansour, 1987). Some of the serotonin analogs, such as dimethyltryptamine, act as activators of the cyclase at high concentrations but are antagonists to the effects of serotonin at lower concentrations. Lysergic acid diethylamide (LSD) is a semisynthetic derivative of the ergot alkaloid, lysergic acid. It is considered in mammalian pharmacology to be a serotonin agonist and produces psychedelic effects in humans at a dose as low as 20 μg. LSD is the most potent activator of the liver fluke adenylyl cyclase ($K_a = 0.045$ μM) but had only 25% of the relative stimulatory efficacy of serotonin. A derivative of LSD, 2-bromo-lysergic acid diethylamide has no stimulatory effect on the cyclase but was found to be the most potent antagonist to serotonin activation of adenylyl cyclase ($K_i = 0.03\mu$M) (Northup & Mansour, 1978).

The discovery that serotonin acts like a hormone to stimulate glycolysis in *F. hepatica* raised the question of which enzyme targets does it influence. Experiments in cell-free extracts showed that serotonin would stimulate glycolysis when glucose, glucose 6-P, or fructose 6-P were used as substrates for glycolysis. However, when fructose 1,6-bisP was the substrate for glycolysis there was almost no difference whether serotonin was added or not (Mansour & Mansour, 1962). The conclusion was that once PFK was bypassed by providing the product of its reaction, fructose 1,6-bisP, the rate of glycolysis was enhanced without stimulation by serotonin. This indicates that PFK is a rate-limiting enzyme in *F. hepatica* and that serotonin increases glycolysis by increasing the activity of PFK. More direct evidence was obtained by demonstrating that either serotonin or cyclic-3',5'-AMP can activate PFK in homogenates of *F. hepatica* (Mansour & Mansour, 1962). The activation of PFK was shown to depend on the presence of ATP, Mg^{2+}, and adenylyl cyclase in the homogenate particles. These experiments suggested the participation of a phosphorylating system that can activate PFK. Verification of regulation of PFK by phosphorylation was shown using PFK from *F. hepatica* that had been purified to homogeneity (Kamemoto & Mansour, 1986; Kamemoto *et al.*, 1987).

Transmembrane Signal Transduction. Molecular entities involved in the transmembrane signaling system of serotonin have been extensively studied in two parasites, *F. hepatica* and *S. mansoni*. The effect of serotonin on glycolysis in these parasites is transmitted across the membrane by coupling the receptors with heterotrimeric guanine nucleotide-binding regulatory proteins (G-proteins). The G-protein is coupled to a very active adenylyl cyclase in the cell membranes. Activation of adenylyl cyclase by serotonin in the parasites is similar to activation of the mammalian cyclase by adrenergic agents (Gilman, 1989). Cyclic AMP released from the activated adenylyl cyclase is available in the cytoplasm for activation of protein kinase, which in turn activates enzymes that control the rate of glycolysis. $G_{s\alpha}$-proteins were identified in adult and intermediate stages of *S. mansoni* (Estey & Mansour, 1988) and in *F. hepatica* (Mansour & Mansour, 1986). The α-subunit of the *S. mansoni* G_s-protein has been cloned (Iltzsch *et al.*, 1992). In spite of the distant phylogenetic relationship between the schistosome and the mammalian host the schistosome $G_{s\alpha}$ is very homologous to the bovine $G_{s\alpha}$ and includes all its major structural features.

Studies done on *F. hepatica* and *S. mansoni* have verified that the same steps for signal transduction in mammals occur in the parasites. Antagonists that compete for serotonin activation of adenylyl cyclase, such as 2-bromo lysergic acid diethylamide and related analogs, exert their inhibitory effect on cyclase activity by competing with serotonin for occupancy of the serotonin receptors (Northup & Mansour, 1978; Estey & Mansour, 1988). It is probable that serotonin receptors and their signal transduction proteins in these flukes are typical of receptors in other trematodes. The serotonin receptors should be further studied as a possible target.

PFK as a Target for Antischistosomal Antimonials

The critical role played by PFK in the regulation of glycolysis and survival of the parasites has also been demonstrated in studies on the mechanism of action of antischistosomal agents. Antimony potassium tartrate and stibophen (sodium antimony III pyrocatechol-2,4-disulfonate) are trivalent antimonials that were the drugs of choice against schistosomiasis for several decades. Both antimonials owe their antischistosomal effect to inhibition of glycolysis in the parasites (Mansour & Bueding, 1954). The antimonials, when present in schistosome cultures *in vitro* at concentrations that may well be achieved *in vivo*, are associated with a reduction in glycolysis and in survival of the parasites (Bueding & Mansour, 1957). *In vitro* and *in vivo* investigations indicate that the inhibition of glycolysis is due to a selective inhibition of schistosome PFK. The PFK of schistosomes is inhibited by the trivalent antimonials at concentrations that do

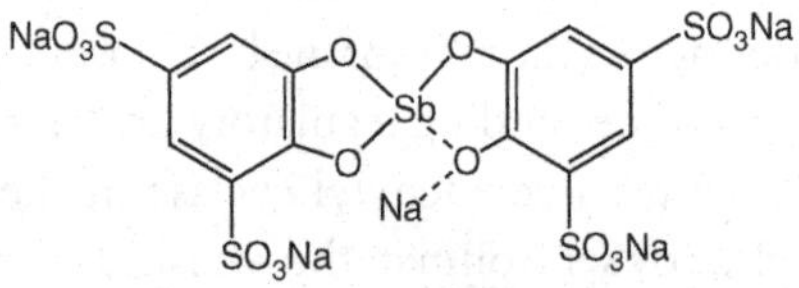

STIBOPHEN

ANTIMONY POTASSIUM TARTRATE

not affect mammalian PFK. Furthermore, in intact parasites, exposure to low concentrations of antimony potassium tartrate results in an accumulation of fructose 6-P, a substrate of PFK, and reduction of fructose 1,6-bisP, a product of the PFK reaction.

Recently, *S. mansoni* PFK was cloned, expressed in insect cells, and purified to homogeneity (Ding, Su, & Mansour, 1994; Su, Mansour, & Mansour, 1996). Recombinant PFK from schistosomes, as well as native PFK from the same parasites, is more sensitive to inhibition by the trivalent antimonials than is mammalian PFK. Stibophen is not as selective an inhibitor of parasite PFK as is antimony potassium tartrate. Inhibition of schistosome PFK is at least partially antagonized by the sulfhydryl protective reagent, dithiothreitol. This indicates involvement of sulfhydryl groups in the selective inhibitory effects of the trivalent antimonials.

Role of Thiol Groups in Phosphofructokinase Activity. Reaction of mammalian PFK with a variety of sulfhydryl reagents was shown to considerably influence its activity and its cooperative kinetics (Younathan, Paetkau, & Lardy, 1968). Enzyme activity is inhibited by monovalent and divalent mercurials and by disulfides. Both PFK from the mammalian host and the parasite are well endowed with cysteine residues. Mammalian PFK has twenty sulfhydryl groups whereas *S. mansoni* has thirteen. In the schistosome enzyme there are eleven cysteines that are unique to the parasite. The mammalian enzyme has four unique cysteines (Ding *et al.*, 1994). Cunningham & Fairlamb (1995) have suggested that the selective inhibition by trivalent antimonials of trypanothione reductase from *Leishmania donovani* could be at least partially due to the formation of monothio-stibane adducts.

Inhibition of PFK from Other Parasites by Antimonials. The selective inhibitory effects of antimonials on parasite PFK is not restricted to the enzyme from schistosomes. Inhibition by antimonials has also been demonstrated in PFKs from the filarial nematodes, *Litomosoides carinii, Dipetalonema witei,* and *Brugia pahangi* and from *Ascaris suum* and the tapeworm *Hymenolepis diminuta* (Saz & Dunbar, 1975). In these parasites stibophen inhibits glycolysis as well as phosphofructokinase activity. Parasite PFK inhibition occurs at concentrations that do not inhibit the mammalian enzyme. In these parasites, the inhibitory effect of stibophen is greater than that of other trivalent antimonials such as antimony potassium tartrate. This selective inhibition of PFK has also been shown in a protozoal parasite, *Trypanosoma rhodesiense* (Jaffe, McCormack, & Meymarian, 1971).

Enzymes Involved in Phosphoenolpyruvate Metabolism

Phosphoenolpyruvate (PEP) is at a bifurcation site in the glycolytic pathway. The fermentation end products of different organisms are determined by the direction their metabolism goes from PEP. In *S. mansoni*, which is a predominately lactate fermenter, pyruvate kinase (PK) catalyzes the main flux of the glycolytic pathway to convert PEP to pyruvate correlated with synthesis of ATP from ADP (see Fig. 3.1). PK in partially purified preparations from either *S. mansoni* or *F. hepatica* is controlled through its allosteric kinetics. This is manifested by the sigmoidal kinetics for the enzyme substrate PEP. Furthermore, the enzyme is inhibited by ATP and activated by fructose 1,6-bisP (Brazier & Jaffe, 1973; Behm & Bryant, 1980). These properties are similar to those of the corresponding mammalian enzyme.

In the case of *Ascaris* muscle and the cestode *Hymenolepis diminuta* the main flux of PEP metabolism is directed toward succinate and fatty acids. *Ascaris* muscle, which produces no appreciable quantities of lactate, has PK activity that is only 4% that of PEP carboxykinase (Bueding & Saz, 1968; Carter & Fairbairn, 1975). In the case of 6-day-old (in the host) *Hymenolepis diminuta* the predominant metabolic product is lactate and, therefore, PEP is directed toward the pyruvic kinase reaction. As the parasites grow older lactate production is decreased and mitochondrial metabolic products (succinate and acetate) increase. This change in PEP metabolism is accompanied by an increase in PEP carboxykinase and a decrease in PK activity.

Mitochondrial Anaerobic Energy-Generating Pathways in Helminths

In mammalian host cells oxidative phosphorylation is the main metabolic pathway by which ATP is formed as electrons are transferred from NADH to O_2 using a series of electron carriers. Parasites that live in a microaerobic environment in the host have acquired an anaerobic form of energy production that does not use molecular oxygen as an electron acceptor. *Ascaris suum* has been the most extensively studied species (Saz, 1981). Many of the anaerobic energy-generating pathways in *Ascaris* have been identified in other parasites such as the trematode *Fasciola hepatica* (Van Hellemond et al., 1996) and the tapeworm *Hymenolepis diminuta* (Scheibel & Saz, 1966). In these organisms glucose or glycogen is degraded anaerobically to PEP via the Embden–Meyerhof reactions (see Fig. 3.1). Subsequently, PEP is directed toward an anaerobic energy yielding pathway (Fig. 3.3). Reactions start in the cytosol where PEP fixes CO_2

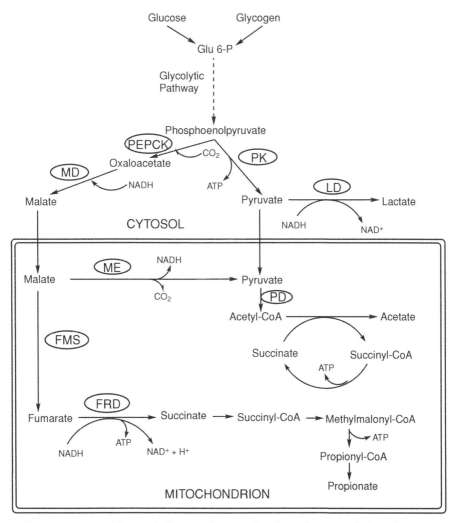

Figure 3.3. Anaerobic metabolism in the mitochondria of parasitic helminths. FMS, fumarase; FRD, fumarate reductase; LD, lactic dehydrogenase; MD, malate dehydrogenase; ME, malic enzyme; PD, pyruvic dehydrogenase; PEPCK, phosphoenolpyruvate carboxykinase; PK, pyruvate kinase. Adapted from (Tielens 1994).

to form oxaloacetate (OAA), a reaction that is catalyzed by PEP carboxykinase. Using NADH formed by the glycolytic pathway, OAA is reduced to malate by malate dehydrogenase. The NAD formed in this reaction serves as an electron acceptor for the triose step in glycolysis to keep an even balance between reduced and oxidized NAD. The malate formed in the cytosol plays the main role in energy production in these parasites. The initial step in this pathway is a translocation of the malate from its initial cytosolic site to the mitochondria. Mitochondria from adult parasites that live in an anaerobic environment lack a tricarboxylic acid cycle and cytochrome oxidase. In the mitochondria of these organisms malate undergoes further major metabolic degradation to produce energy by a dismutation reaction. Half of the malate is oxidized via pyruvate to acetate. First malate is oxidized to pyruvate and CO_2 by the malic enzyme. Then pyruvate is converted to acetylCoA by pyruvate dehydrogenase and on to acetate via a succinate–succinal CoA cycle. The other half of the malate in the mitochondria is reduced via fumarate to succinate. Malate is first converted to fumarate by fumarase and then reduced to succinate by fumarate reductase, an NADH-utilizing enzyme. Reduction of fumarate to succinate is coupled to an electron transport–linked phosphorylation of ADP at complex I of the respiratory chain, rhodoquinone and fumarate reductase. Some parasites including *Fasciola hepatica* metabolize succinate further to propionate. The anaerobic degradation of malate to acetate and propionate is accompanied by the formation of three molecules of ATP. Rhodoquinone appears to be an essential component of electron transport–associated phosphorylation in all anaerobic parasitic helminths that reduce fumarate. Complex I is composed of a flavin-containing subunit, a subunit that contains three iron-sulfur clusters, and two hydrophobic cytochrome b-containing subunits (Saz, 1981; Tielens, 1994; Komuniecki & Komuniecki, 1995; Tielens & Van Hellemond, 1998).

The anaerobic mitochondria of *Ascaris* and related parasites are able to synthesize 2-methyl-branched-chain-fatty-acids-enoyl-CoAs. This is carried out through a series of condensation reactions of acetyl CoA and propionyl CoA, forming two branched-chain fatty acids, 2-methylbutyrate and 2-methylvalerate. Reduction of these fatty acids catalyzed by 2-methyl-branched-chain-enoyl-CoA reductase, as in the case of fumarate reductase, is coupled to site I electron transfer flavoprotein, rhodoquinone, which is associated with ADP phosphorylation. The enzyme has been cloned and characterized from *Ascaris suum* (Duran *et al.*, 1993).

Obviously, there are remarkable differences between the biochemical machinery in these anaerobic parasites and equivalent reactions in the mammalian host that would be of great interest to those who are seeking selective targets for prospective new chemotherapeutic agents. A brief discussion of some of the enzyme systems involved in these reactions is given here. Some of the enzymes participating in anaerobic mitochondrial energy metabolism are being extensively studied because of their critical role in the regulation of these pathways.

Phosphoenolpyruvate Carboxykinase (PEPCK)

Although PEPCK (see Fig. 3.3) is present in parasites as well as in their mammalian hosts it serves different functions. In the host PEPCK catalyzes the reaction from oxaloacetate to make PEP and CO_2, part of the gluconeogenesis pathway. Parasite PEPCK catalyzes the reverse reaction to form oxaloacetate, the precursor of malate, which is metabolized anaerobically by the mitochondria. This functional difference is related to significant differences in the molecular properties of the parasite PEPCK. Studies on PEPCK purified from *Ascaris suum* revealed important kinetic characteristics of the enzyme that are related to the active sites. These include a special role in the *Ascaris* muscle for Mn^{2+}, which controls enzyme activity by increasing the affinity of PEP and the nucleotide (GTP) to the active site (Rohrer, Saz, & Nowak, 1986). Both *Ascaris* and *Haemonchus* enzymes have been cloned. Sequence analysis of both cDNAs indicate that the putative active site for GTP is the same when compared with other PEPCKs. However, putative sites for PEP binding were more conserved between *A. suum* and *H. contortus* but less conserved when compared with that of the rat, chicken, and *Drosophila*. These differences between the mammalian and the parasite enzymes support the belief that PEPCK should be further investigated as a possible target for selective chemotherapeutic agents (Klein *et al.* 1992; Geary *et al.*, 1993).

Fumarate Reductase

Fumarate reductase (Fig. 3.3), a key parasite enzyme for energy production, catalyzes the reverse of the reaction of mammalian succinic dehydrogenase in the Krebs cycle. This reflects significant differences between the two enzymes. Also, fumarate reductase is linked to a respiratory chain that is unique to these parasites. In the mammalian respiratory chain the quinone electron carrier is ubiquinone. However, in adult *Ascaris* and *F. hepatica*, anaerobic metabolism utilizes fumarate reduction coupled to rhodoquinone (Van Hellemond *et al.*, 1996).

Pyruvate Dehydrogenase Complex (PDC)

Another enzyme that has received serious attention is the pyruvate dehydrogenase complex. In mammals PDC is a group of mitochondrial enzymes as well as other structural proteins. It is, therefore, always referred to as a "complex." The main enzymes in the PDC catalyze the oxidative decarboxylation of pyruvate to form acetyl CoA, which is incorporated into the citric acid cycle for oxidative reactions:

$$\text{pyruvate} + \text{CoA} + \text{NAD} \longrightarrow \text{acetyl CoA} + CO_2 + \text{NADH.}$$

Ascaris early larval stages have access to an aerobic environment and they utilize aerobic metabolism. PDC in these early larval stages has a function like that of its mammalian counterpart and supplies acetyl CoA to the tricarboxylic cycle for oxidation. Adult *Ascaris* have a predominantly anaerobic fermentative metabolism and their PDC occupies a critical position in supplying their mitochondrial machinery with acetyl CoA and high levels of NADH. The *Ascaris* enzyme complex contains the same three enzymes as mammalian PDC: pyruvate dehydrogenase, dihydrolipoyl transacetylase, and dihydrolipoyl dehydrogenase. Anaerobic metabolism of malate by the malic enzyme via pyruvate and PDC catalyzes the formation of acetyl CoA, CO_2, and NADH. Pyruvate dehydrogenase activity is controlled by phosphorylation–dephosphorylation enzyme systems (Komuniecki, Wack, & Coulson, 1983). Enzyme activity is inhibited when the enzyme is phosphorylated by a special protein kinase that is tightly associated with PDC. Also, there is a special pyruvate dehydrogenase phosphatase connected with PDC. Regulation of PDC activity in the *Ascaris* anaerobic mitochondria appears to be different from that of their mammalian hosts (Song & Komuniecki, 1994). The interaction between different components of *Ascaris* PDC regulates this multiprotein complex. An important protein that is thought to have binding sites for dihydrolipoyl dehydrogenase was found to be missing from the *Ascaris* PDC (Komuniecki *et al.*, 1992).

Anaerobic Mitochondrial Metabolism as Possible Targets for Antiparasitic Agents

Early research on the pathways of anaerobic mitochondrial phosphorylation systems in parasitic helminths gave indications that some of the pathways are quite different from those in the mammalian host and that there would be a good chance of finding inhibitors that affect the parasite pathway selectively. There are a few reports on the mechanism of action of anthelmintics that give credence to this supposition.

Using *Ascaris* muscle mitochondria Saz (1971) was able to establish *in vitro* conditions to determine anaerobic malate-dependent phosphorylation. This was done by measuring the incorporation of inorganic phosphate [$^{32}P_i$] into the organic fraction of the *Ascaris* mitochondrial system. This represents ATP generation and is an ideal biochemical system for measuring malate dismutation as a function of inorganic phosphate esterification to ATP. In this system

DITHIAZANINE

dinitrophenol, a classical uncoupler of oxidative phosphorylation, was shown to inhibit the anaerobic phosphorylation in the *Ascaris* mitochondrial system. Dithiazanine, a broad spectrum antinematodal agent, was also shown to cause ~50% inhibition of the anaerobic phosphorylation system at micromolar concentration. The high sensitivity of the *Ascaris* mitochondrial system to dithiazanine suggests that the inhibitory effect on the anaerobic phosphorylation may be related to the antinematodal effects of the drug.

The cestode *Hymenolepis diminuta*, an intestinal parasite of rats and other rodents, is used as a model organism for studies on the mechanism of action of anticestodal agents. The parasite shares with *Ascaris* many of the anaerobic mitochondrial phosphorylation reactions that are associated with electron transport. The major metabolic product of carbohydrate fermentation in *H. diminuta* is succinate. Several compounds with known anticestodal effects *in vivo* were shown to cause a marked decrease in the activity of reactions involved in inorganic phosphate incorporation into ATP under anaerobic conditions in the mitochondria (Scheibel, Saz, & Bueding, 1968). These compounds include desaspidin (an active principle from the oleoresin of aspidium), dichlorophen, and niclosamide (Yomesan). For a review of the antiparasitic uses of these compounds see Andrews & Bonse (1986). The experimental evidence indicates that these anthelmintic agents at very low concentrations act as uncouplers of the cestodal mitochondrial electron transport system. This is similar to the effect of 2,4-dinitrophenol on the oxidative phosphorylation in mammalian mitochondria. More evidence is needed to identify the primary site of action of these anticestodal agents that are responsible for inhibition of mitochondrial anaerobic phosphorylation.

Studies by Van den Bossche on the effect of tetramisole, a broad spectrum antinematodal agent, showed a selective inhibitory effect against fumarate reductase of *Ascaris suum* and other related nematodes (Van den Bossche & Janssen, 1969). Tetramisole is a mixture of a levorotatory isomer and the corresponding dextro isomer. The levorotatory isomer is a highly superior inhibitor of the enzyme when compared to the dextro isomer. Tetramisole was a more potent inhibitor when the parasite enzyme was measured as fumarate reductase compared to the reverse reaction, succinate dehydrogenase in the host. However,

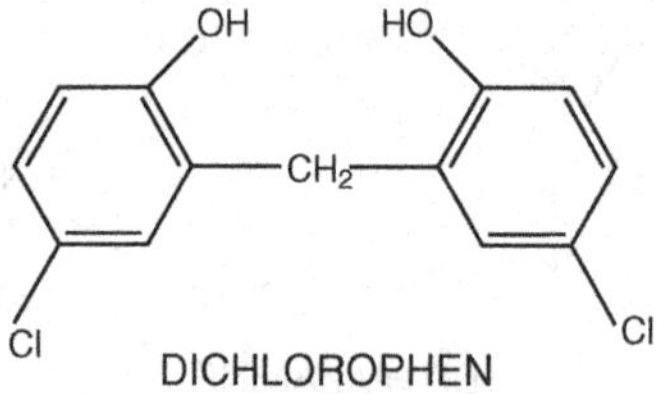

DICHLOROPHEN

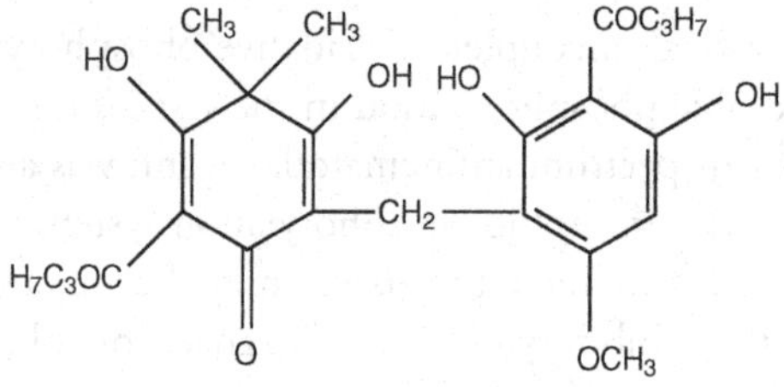

DESASPIDIN

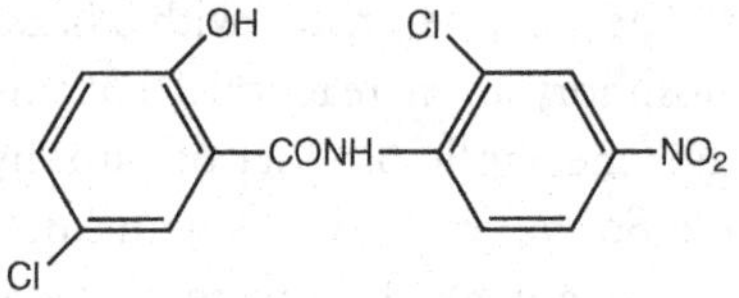

NICLOSAMIDE

tetramisole had no effect on succinate oxidation in the pigeon heart muscle mitochondrial system and only a minor inhibitory effect at drug concentrations fifty times higher than those necessary to inhibit fumarate reductase in the nematode. The apparent selective inhibitory effect of tetramisole is not surprising when it is considered that the parasite enzyme favors catalysis of the reverse reaction of the host's succinate dehydrogenase.

Bithionol (2, 2′-thiobis-dichlorophenol) is another antinematodal agent that appears to indirectly affect the NADH-fumarate reductase system of *Ascaris* muscle. The inhibitory effect was demonstrated in a partially purified and solubilized mitochondrial enzyme from *Ascaris suum*. In this system the enzyme was separated from rhodoquinone (a benzoquinone) of the electron transport chain. Recovery of enzyme activity was dependent on its reconstitution with

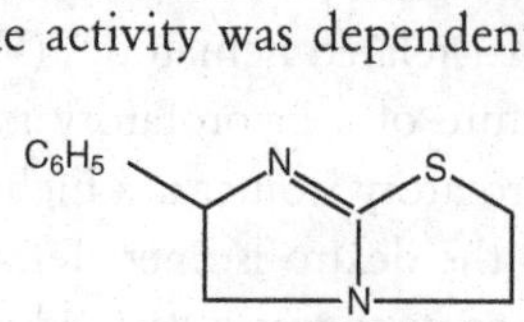

TETRAMISOLE

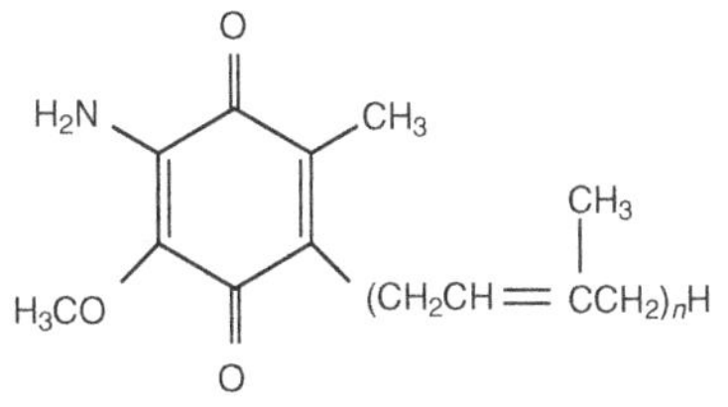

RHODOQUINONE

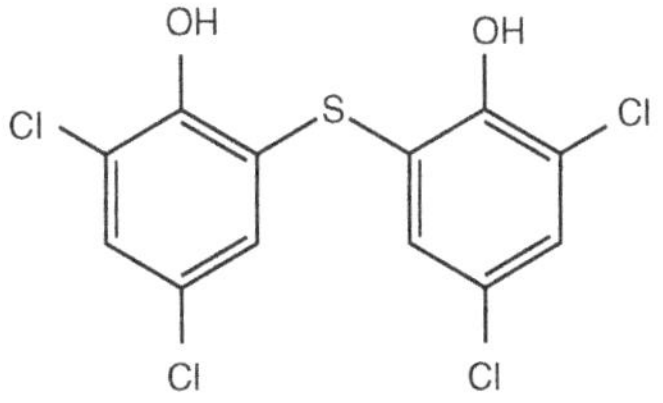

BITHIONOL

rhodoquinone. Bithionol was found to be a potent inhibitor of rhodoquinone-activated NADH-fumarate reductase. The inhibition was competitive with respect to the concentration of rhodoquinone. The results indicate that the anthelmintic action of bithionol is due to competition between two chemically related compounds, bithionol and rhodoquinone, where the outcome is inhibition of the electron transfer to fumarate and inhibition of anaerobic energy metabolism in the parasites (Ikuma, Makimura, & Murakoshi, 1993).

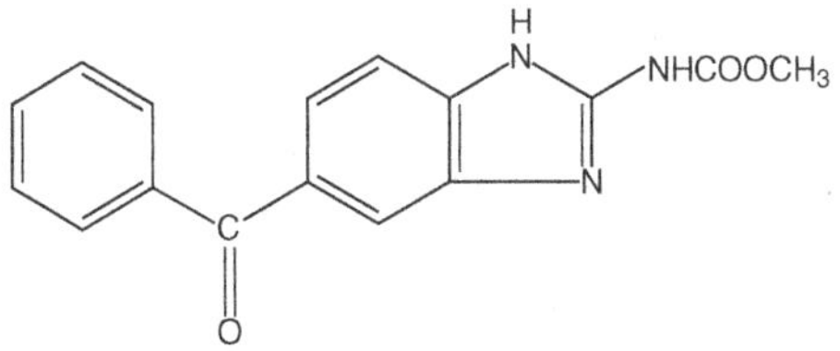

MEBENDAZOLE

Mebendazole, a benzimidazole derivative with a potent anthelmintic effect against adult and larval stages of *A. lumbricoides,* was shown to be a potent inhibitor of both soluble and mitochondrial partially purified preparations of malic dehydrogenase. This effect appears to be more specific for mebendazole when compared to other benzimidazole anthelmintics such as albendazole, parabendazole, and thiabendazole. The inhibitory effect is more selective against malic dehydrogenase from *Ascaris* when compared to the same enzyme from the trematode *Fasciola hepatica* and the cestode *Moniezia expansa*. It has been suggested that inhibition of the malic dehydrogenase contributes to a decline in the parasite energy reserves (Tejada *et al.*, 1987).

It can be surmised that anaerobic mitochondrial phosphorylation in intestinal nematodes and some helminths that use the same anaerobic mitochondrial electron transport system is sufficiently different from homologous anaerobic phosphorylation reactions in the host to be amenable to inhibition by chemical agents that do not significantly affect the host mitochondrial systems. Further experimentation on the components of mitochondrial phosphorylation enzyme systems and their sensitivity to different inhibitors would be well justified.

REFERENCES

Andrews, P. & Bonse, G. (1986). Chemistry of anticestodal gents. In W. Campbell & R. Rew (Eds.), *Chemostherapy of Parasitic Diseases* (pp. 447–456). New York: Plenum Press.

Behm, C. A. & Bryant, C. (1980). Regulatory properties of a partially purified preparation of pyruvate kinase from *Fasciola hepatica*. *Int J Parasitol, 10*(2), 107–114.

Brazier, J. B. & Jaffe, J. J. (1973). Two types of pyruvate kinase in schistosomes and filariae. *Comp Biochem Physiol [B], 44*(1), 145–155.

Bueding, E. (1950). Carbohydrate metabolism of *Schistosoma mansoni*. *J Gen Physiol, 33*, 475–495.

Bueding, E. & Fisher, J. (1982). Metabolic requirements of schistosomes. *J Parasitol, 68*(2), 208–212.

Bueding, E. & Mansour, J. (1957). The relationship between inhibition of phosphofructokinase activity and the mode of action of trivalent organic antimonials on *Schistosoma mansoni*. *Brit J Pharmacol, 12*, 159–165.

Bueding, E. & Saz, H. J. (1968). Pyruvate kinase and phosphoenolpyruvate carboxykinase activities of *Ascaris* muscle, *Hymenolepis diminuta* and *Schistosoma mansoni*. *Comp Biochem Physiol, 24*(2), 511–518.

Carter, C. E. & Fairbairn, D. (1975). Multienzymic nature of pyruvate kinase during development of *Hymenolepis diminuta* (Cestoda). *J Exp Zool, 194*(2), 439–448.

Coles, G. (1973). The metabolism of schistosomes: A review. *Int J Biochem, 4*, 319–337.

Cunningham, M. L. & Fairlamb, A. H. (1995). Trypanothione reductase from *Leishmania donovani*. Purification, characterisation and inhibition by trivalent antimonials. *Eur J Biochem, 230*(2), 460–468.

Ding, J., Su, J. G. & Mansour, T. E. (1994). Cloning and characterization of a cDNA encoding phosphofructokinase from *Schistosoma mansoni*. *Mol Biochem Parasitol, 66*(1), 105–110.

Duran, E., Komuniecki, R. W., Komuniecki, P. R., Wheelock, M. J., Klingbeil, M. M., Ma, Y. C. & Johnson, K. R. (1993). Characterization of cDNA clones for the 2-methyl branched-chain enoyl-CoA reductase. An enzyme involved in branched-chain fatty acid synthesis in anaerobic mitochondria of the parasitic nematode *Ascaris suum*. *J Biol Chem, 268*(30), 22391–22396.

Estey, S. J. & Mansour, T. E. (1987). Nature of serotonin-activated adenylate cyclase during development of *Schistosoma mansoni*. *Mol Biochem Parasitol, 26*(1–2), 47–59.

Estey, S. J. & Mansour, T. E. (1988). GTP binding regulatory proteins of adenylate cyclase in *Schistosoma mansoni* at different stages of development. *Mol Biochem Parasitol, 30*(1), 67–75.

Geary, T. G., Winterrowd, C. A., Alexander-Bowman, S. J., Favreau, M. A., Nulf, S. C. & Klein, R. D. (1993). *Ascaris suum*: Cloning of a cDNA encoding phosphoenolpyruvate carboxykinase. *Exp Parasitol, 77*(2), 155–161.

Gilman, A. G. (1989). The Albert Lasker Medical Awards. G proteins and regulation of adenylyl cyclase. *J Am Med Soc, 262*(13), 1819–1825.

Ikuma, K., Makimura, M. & Murakoshi, Y. (1993). Inhibitory effect of bithionol on NADH-fumarate reductase in ascarides. *Yakugaku Zasshi, 113*(9), 663–669.

Iltzsch, M. H., Bieber, D., Vijayasarathy, S., Webster, P., Zurita, M., Ding, J. & Mansour, T. E. (1992). Cloning and characterization of a cDNA coding for the alpha-subunit of a stimulatory G protein from *Schistosoma mansoni. J Biol Chem, 267*(20), 14504–14508.

Jaffe, J. J., McCormack, J. J. & Meymarian, E. (1971). Phosphofructokinase activity in particulate-free extracts of *Trypanosoma (Trypanozoon) rhodesiense. Comp Biochem Physiol [B], 39*(4), 775–788.

Kamemoto, E. S. & Mansour, T. E. (1986). Phosphofructokinase in the liver fluke *Fasciola hepatica*. Purification and kinetic changes by phosphorylation. *J Biol Chem, 261*(9), 4346–4351.

Kamemoto, E. S., Iltzsch, M. H., Lan, L. & Mansour, T. E. (1987). Phosphofructokinase from *Fasciola hepatica*: Activation by phosphorylation and other regulatory properties distinct from the mammalian enzyme. *Arch Biochem Biophys, 258*(1), 101–111.

Kemp, R. G. (1969). Allosteric properties of muscle phosphofructokinase. II. Kinetics of native and thiol-modified enzyme. *Biochemistry, 8*(11), 4490–4496.

Klein, R. D., Winterrowd, C. A., Hatzenbuhler, N. T., Shea, M. H., Favreau, M. A., Nulf, S. C. & Geary, T. G. (1992). Cloning of a cDNA encoding phosphoenolpyruvate carboxykinase from *Haemonchus contortus. Mol Biochem Parasitol, 50*(2), 285–294.

Komuniecki, R. & Komuniecki, P. (1995). Aerobic–anaerobic transitions in energy metabolism during the development of the parasitic nenatode *Ascaris suum*. In J. C. Boothroyd & R. Komuniecki (Eds.) *Molecular Approaches to Parasitology* (pp. 109–121). New York: Wiley-Liss.

Komuniecki, R., Wack, M. & Coulson, M. (1983). Regulation of the *Ascaris suum* pyruvate dehydrogenase complex by phosphorylation and dephosphorylation. *Mol Biochem Parasitol, 8*(2), 165–176.

Komuniecki, R., Rhee, R., Bhat, D., Duran, E., Sidawy, E. & Song, H. (1992). The pyruvate dehydrogenase complex from the parasitic nematode *Ascaris suum*: Novel subunit composition and domain structure of the dihydrolipoyl transacetylase component. *Arch Biochem Biophys, 296*(1), 115–121.

Kulkarni, G., Rao, G. S., Srinivasan, N. G., Hofer, H. W., Yuan, P. M. & Harris, B. G. (1987). *Ascaris suum* phosphofructokinase. Phosphorylation by protein kinase and sequence of the phosphopeptide. *J Biol Chem, 262*(1), 32–34.

Mahrenholz, A. M., Hefta, S. A. & Mansour, T. E. (1991). Phosphofructokinase from *Fasciola hepatica*: Sequence of the cAMP-dependent protein kinase phosphorylation site. *Arch Biochem Biophys, 288*(2), 463–467.

Mansour, J. M. & Mansour, T. E. (1986). GTP-binding proteins associated with serotonin-activated adenylate cyclase in *Fasciola hepatica. Mol Biochem Parasitol, 21*(2), 139–149.

Mansour, T. (1957). The effect of lysergic acid diethylamide, 5-hydroxytryptamine, and related compounds on the liver fluke *Fasciola hepatica*. *Brit J Pharmacol, 12*, 406–409.

Mansour, T. & Bueding, E. (1954). The actions of antimonials on glycolytic enzymes of *Schistosoma mansoni*. *Brit J Pharmacol Chemother, 9*, 459–462.

Mansour, T. & Mansour, J. (1962). Effects of serotonin (5-hydroxytrypatmine) and adenosine $3',5'$-phosphate on phosphofructokinase from the liver fluke *Fasciola hepatica*. *J Biol Chem, 237*, 629–634.

Mansour, T., Sutherland, E., Rall, T. & Bueding, E. (1960). The effect of serotonin (5-hydroxytryptamine) on the formation of adenosine $3',5'$-phosphate by tissue particles from the liver fluke, *Fasciola hepatica*. *J Biol Chem, 235*, 466–470.

Mansour, T. E. (1972). Phosphofructokinase. *Curr Top Cell Regul, 5*, 1–46.

Mettrick, D. F. & Cho, C. H. (1981). Migration of *Hymenolepis diminuta* (Cestoda) and changes in 5-HT (serotonin) levels in the rat host following parenteral and oral 5-HT administration. *Can J Physiol Pharmacol, 59*(3), 281–286.

Northup, J. K. & Mansour, T. E. (1978). Adenylate cyclase from *Fasciola hepatica*. 1. Ligand specificity of adenylate cyclase-coupled serotonin receptors. *Mol Pharmacol, 14*(5), 804–819.

Pasteur, L. (1861). Experiences et ues nouvelles sur la nature des fermentations. *C R Acad Sci Paris, 52*, 1260–1264.

Rohrer, S. P., Saz, H. J. & Nowak, T. (1986). Purification and characterization of phosphoenolpyruvate carboxykinase from the parasitic helminth *Ascaris suum*. *J Biol Chem, 261*(28), 13049–13055.

Saz, H. J. (1971). Anaerobic phosphorylation in *Ascaris* mitochondria and the effects of anthelmintics. *Comp Biochem Physiol [B], 39*(3), 627–637.

Saz, H. J. (1981). Energy metabolisms of parasitic helminths: Adaptations to parasitism. *Annu Rev Physiol, 43*, 323–341.

Saz, H. J. & Dunbar, G. A. (1975). The effects of stibophen on phosphofructokinases and aldolases of adult filariids. *J Parasitol, 61*(5), 794–801.

Scheibel, L. W. & Saz, H. J. (1966). The pathway for anaerobic carbohydrate dissimilation in *Hymenolepis diminuta*. *Comp Biochem Physiol, 18*(1), 151–162.

Scheibel, L. W., Saz, H. J. & Bueding, E. (1968). The anaerobic incorporation of ^{32}P into adenosine triphosphate by *Hymenolepis diminuta*. *J Biol Chem, 243*(9), 2229–2235.

Setlow, B. & Mansour, T. E. (1970). Studies on heart phosphofructokinase. Nature of the enzyme desensitized to allosteric control by photo-oxidation and by acylation with ethoxyformic anhydride. *J Biol Chem, 245*(21), 5524–5533.

Shoemaker, C. B., Reynolds, S. R., Wei, G., Tielens, A. G. & Harn, D. A. (1995). *Schistosoma mansoni* hexokinase: cDNA cloning and immunogenicity studies. *Exp Parasitol, 80*(1), 36–45.

Silverman, M. (1991). Structure and function of hexose transporters. *Annu Rev Biochem, 60*, 757–794.

Skelly, P. J. & Shoemaker, C. B. (1996). Rapid appearance and asymmetric distribution of glucose transporter SGTP4 at the apical surface of intramammalian-stage *Schistosoma mansoni*. *Proc Natl Acad Sci USA, 93*(8), 3642–3646.

Skelly, P. J., Kim, J. W., Cunningham, J. & Shoemaker, C. B. (1994). Cloning, characterization, and functional expression of cDNAs encoding glucose transporter proteins from the human parasite *Schistosoma mansoni*. *J Biol Chem, 269*(6), 4247–4253.

Smyth, J. D. (1994). *Introduction to Animal Parasitoloty* (3rd ed.). Cambridge, England: Cambridge University Press.

Song, H. & Komuniecki, R. (1994). Novel regulation of pyruvate dehydrogenase phosphatase purified from anaerobic muscle mitochondria of the adult parasitic nematode, *Ascaris suum*. *J Biol Chem, 269*(50), 31573–31578.

Su, J.-G., Mansour, J. & Mansour, T. (1996). Purification, kinetics and inhibition by antimonials of recombinant phosphofructokinase from *Schistosoma mansoni*. *Mol Biochem Parasitol, 81*, 171–178.

Tejada, P., Sanchez-Moreno, M., Monteoliva, M. & Gomez-Banqueri, H. (1987). Inhibition of malate dehydrogenase enzymes by benzimidazole anthelmintics. *Vet Parasitol, 24*(3–4), 269–274.

Tielens, A. (1994). Energy generation in parasitic helminths. *Parasitol Today, 10*, 346–352.

Tielens, A. & Van Hellemond, J. (1998). The electron transport chain in anaerobically functioning eukaryotes. *Biochim Biophys Acta, 1365*, 71–78.

Tielens, A. G., Houweling, M. & Van den Bergh, S. G. (1985). The effect of 5-thioglucose on the energy metabolism of *Schistosoma mansoni* in vitro. *Biochem Pharmacol, 34*(18), 3369–3373.

Tielens, A. G., Horemans, A. M., Dunnewijk, R., van der Meer, P. & van den Bergh, S. G. (1992). The facultative anaerobic energy metabolism of *Schistosoma mansoni* sporocysts. *Mol Biochem Parasitol, 56*(1), 49–57.

Tielens, A. G., van den Heuvel, J. M., van Mazijk, H. J., Wilson, J. E. & Shoemaker, C. B. (1994). The 50-kDa glucose 6-phosphate-sensitive hexokinase of *Schistosoma mansoni*. *J Biol Chem, 269*(40), 24736–24741.

Van den Bossche, H. & Janssen, P. A. (1969). The biochemical mechanism of action of the antinematodal drug tetramisole. *Biochem Pharmacol, 18*(1), 35–42.

Van Hellemond, J. J., Luijten, M., Flesch, F. M., Gaasenbeek, C. P. & Tielens, A. G. (1996). Rhodoquinone is synthesized de novo by *Fasciola hepatica*. *Mol Biochem Parasitol, 82*(2), 217–226.

van Oordt, B. E., van den Heuvel, J. M., Tielens, A. G. & van den Bergh, S. G. (1985). The energy production of the adult *Schistosoma mansoni* is for a large part aerobic. *Mol Biochem Parasitol, 16*(2), 117–126.

Whistler, R. L. & Lake, W. C. (1972). Inhibition of cellular transport processes by 5-thio-D-glucopyranose. *Biochem J, 130*(4), 919–925.

Younathan, E. S., Paetkau, V. & Lardy, H. A. (1968). Rabbit muscle phosphofructokinase. Reactivity and function of thiol groups. *J Biol Chem, 243*(7), 1603–1608.

Zhong, C., Skelly, P. J., Leaffer, D., Cohn, R. G., Caulfield, J. P. & Shoemaker, C. B. (1995). Immunolocalization of a *Schistosoma mansoni* facilitated diffusion glucose transporter to the basal, but not the apical, membranes of the surface syncytium. *Parasitology, 110*(Pt 4), 383–394.

4

Antimalarial Agents and Their Targets

Before the basic principles of modern chemotherapy were established in the early part of the twentieth century large numbers of the known antiparasitic agents were adapted from ancient remedies. The rationale for using these agents was based on subjective observations of alleviating symptoms of the disease or, in the case of intestinal parasitic worms, evacuation of the parasites from patients. Once the biology of the parasites' life cycle was established and methods were devised to evaluate the effectiveness of the therapeutic agents, a more scientific approach was taken to assess the mechanism of action of these drugs using *in vitro* cultures or experimental animals.

Malaria remains the most challenging disease that afflicts humankind, both in prevalence and in the morbidity and mortality it causes. It is estimated that 300–500 million people in developing countries have acute infection. Of these, 1.5–2.7 million, mostly children, die every year. The disease is characterized by shaking chills, relapsing fever, mental or physical exhaustion, splenomegaly, and anemia. The disease is caused by four species of the genus *Plasmodium* (*P. falciparum, P. vivax, P. malaria,* and *P. ovale*). *P. falciparum* causes the most serious form of the disease. The vector for the parasite is a female mosquito of the genus *Anopheles.*

The infective stage of malaria, the sporozoite, is injected by the mosquito from its salivary gland while it is feeding on mammalian blood (Fig. 4.1). After about 1 hour of circulating in the mammalian bloodstream the sporozoites enter the hepatic microcirculation and then move into the hepatocytes. Development of the parasite in the hepatocytes lasts for different durations (5–13 days) depending on the type of malaria. This stage is termed the exoerythrocytic schizogony when the parasites multiply rapidly, eventually forming thousands of merozoites. These are released into the bloodstream and enter the erythrocytes through specific receptors, thus establishing the erythrocytic cycle. There are important differences between species as to how the hepatic stage ends. In

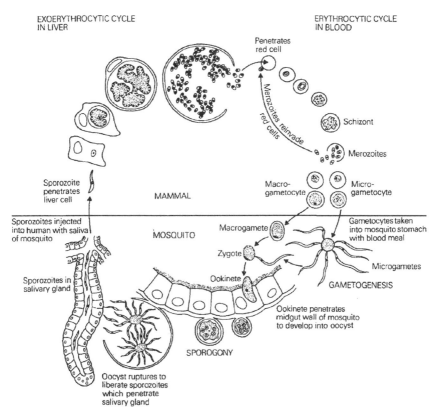

Figure 4.1. The life cycle of *Plasmodium spp.* in mammals and mosquitoes. Sporozoites are injected directly into the bloodstream from the salivary glands of a mosquito. The sporozoites enter liver cells where they begin a phase of multiplication called "exoery-throcytic schizogony," during which thousands of uninucleate merozoites are formed. These enter red blood cells in which they undergo a second phase of multiplication or erythrocytic schizogony. These merozoites invade new red blood cells and the cycle may be repeated many times. Some of the merozoites are capable of developing into sexual stages or gametocytes. These are taken up by a mosquito. In the gut of the mosquito microgametes are produced and these fertilize the macrogametes and the resulting zygote bores through the gut wall to lie on the outer surface, where it forms an oocyst. Within the oocyst a third stage of multiplication occurs, resulting in the formation of sporozoites that enter the salivary glands of the mosquito. After K. Vickerman and F.E.G. Cox, 1967, *The Protozoa*, John Murray, London (with permission) and 1993, *Modern Parasitology*, 2nd ed., F.E.G. Cox, Ed., Blackwell, Oxford.

P. falciparum and *P. malariae* the liver stage terminates at the beginning of the erythrocytic stage. In *P. vivax* and *P. ovale* the liver forms persist, differentiating into stages termed hypnozoites. The dormant hypnozoites remain in the hepatocytes for some time. They undergo schizogony, causing relapse of the disease when the red cells are invaded. The merozoite stage in the erythrocyte grows and develops into a ring-shaped trophozoite and then to multinucleated erythrocytic schizonts. These multiply (8–12 per merozoite) and develop in the infected erythrocytes until they destroy the erythrocytes and are released into the bloodstream to continue the infectious cycle by invading more erythrocytes. After several erythrocytic cycles some merozoites develop into female and male gametocytes, which are transmitted to the stomach of the mosquito as it sucks blood from the mammalian host. In the mosquito stomach fertilization occurs to form zygotes, which develop into ookinetes, which penetrate the stomach wall and multiply, forming a growing oocyst. This stage eventually will rupture, releasing sporozoites that migrate to and accumulate in the salivary gland of the mosquito ready for infecting the next victim.

Chemotherapy of malaria must be targeted not only to the species of the parasites but to a specific stage of parasite development. For example, drugs that are effective against the erythrocytic stages are not effective against the hepatic stages.

Quinine and Quinidine

Both quinine and quinidine are alkaloids from the cinchona trees that grow in South America and the East Indies. The medicinal use of cinchona bark to treat feverish conditions was known to the natives of Peru long before the malaria parasite was recognized. Franciscan monks were the first Europeans to describe the effect of decoctions of the cinchona bark against recurring fevers. Both quinine and quinidine are quinoline methanol derivatives that have been isolated from the natural cinchona bark. They can also be synthesized by a complicated and expensive chemical procedure. Quinine is a mixture of the d- and l-isomers. Quinidine is the pure d-isomer and is more effective as an antimalarial agent. Until the end of World War II, quinine was the main antimalarial

QUININE

agent. Its importance subsided when new synthetic antimalarial agents such as chloroquine were introduced to the therapeutic armamentarium. It was not until the late 1970s that the therapeutic use of quinine was reintroduced when chloroquine-resistant malaria became prevalent. Resistance to quinine or quinidine is uncommon.

The antimalarial effect of quinine and quinidine is restricted to erythrocytic stages of all species of human malaria. Currently, quinine and quinidine are usually used only against chloroquine-resistant *P. falciparum* or the erythrocytic stages of the other human malarias. Because of its availability in preparations for parenteral use quinidine is given as a substitute for quinine when, clinically, parenteral administration is preferred. The mechanism of action of these quinoline derivatives is discussed with chloroquine.

Chloroquine

Chloroquine, a 4-aminoquinoline, was synthesized in 1934 by the German firm of Bayer AG. Chloroquine was introduced as an antimalarial agent as the result of an extensive research program in the United States toward the end of the Second World War. The chlorine at the 7-position of chloroquine is necessary for its antiplasmodial effect. All stages in the parasite's life cycle that digest hemoglobin are sensitive to its effect. The drug is also effective against the gametocytes of *P. vivax*, *P. ovale*, and *P. malariae*. Its gametocidal effect in *P. falciparum* is only against immature forms of that stage. It has no effect against the sporozoites and primary and secondary exoerythrocytic stages of all strains. It also has no effect against the latent liver forms of *P. vivax* or *P. ovale*. Because chloroquine is a safe drug (particularly for children and pregnant women) and is effective against the four types of malaria that infect humans, it remained the drug of choice for more than forty years. Amodiaquine is a congener of chloroquine that shares its antimalarial stage specificities, but it is not frequently used.

CHLOROQUINE

Concentration of Chloroquine in *Plasmodium* Food Vacuoles

The mechanism of the antimalarial action of chloroquine has been the subject of investigations almost since its introduction as a therapeutic agent. Erythrocytes infected with malaria are known to concentrate chloroquine as much as several hundred times when compared to normal erythrocytes (Macomber, O' Brien, & Hahn, 1966). Furthermore, erythrocytes infected with chloroquine-resistant parasites are less able to concentrate chloroquine. The drug accumulates primarily within the acid food vacuoles of the erythrocytic stage of the malaria parasite. The malaria food vacuoles are analogous to mammalian lysosomes. The vacuoles are cellular organelles that contain many degrading enzymes, including proteases for digestion of ingested red cell hemoglobin. The pH of the food vacuoles is estimated at 5.0–5.4 (Krogstad & Schlesinger, 1986), which favors optimal activity of these hydrolytic enzymes, including the proteases. Chemically speaking, chloroquine is a weak base with two protonation sites with alkaline pK_a values of 10.2 and 8.3. Red cell interior and parasite cytoplasm pH values were estimated to be 7.0–7.2 and 6.8–7.0 respectively. This establishes a pH gradient between the extracellular space and the acidic food vacuole of the parasite. As a weak base chloroquine readily crosses the membranes of the erythrocytes and digestive vacuoles in the unprotonated (uncharged) form but poorly in the protonated (charged) form. As soon as chloroquine enters the acidic food vacuoles it is rapidly converted to the protonated form. This has the effect of trapping the membrane-impermeable protonated species of the drug. The total concentration of chloroquine (protonated and unprotonated) increases as additional unprotonated chloroquine (a weak base) moves into the vacuole. A concentration gradient for the inward movement of unprotonated drug is maintained until equilibrium is achieved (Krogstad & Schlesinger, 1987). Based on this process of ion trapping, chloroquine accumulates in the malaria parasite acidic food vacuoles. The physiochemical mechanism for accumulation of chloroquine in the food vacuole does not appear to be the sole reason for high accumulation of chloroquine in the vacuoles. Krogstad also reported that the parasite acidic vacuoles accumulate >1000-fold more chloroquine than the predicted amounts based on the accumulation of the diprotonated base in the food vacuoles. Chloroquine and its congeners (mefloquine, quinine, and quinidine) could reach millimolar concentrations in the digestive vacuole of *P. falciparum* from nanomolar concentrations in the plasma. Thus, the parasite vacuoles are capable of greater chloroquine accumulation than the mammalian acid liposomes (Krogstad & Schlesinger, 1987). In addition, chloroquine-resistant malaria parasites are capable of accumulating the weak base in the food vacuoles at a level intermediate between those of susceptible parasites and mammalian cells. Additional chloroquine-concentrating mechanisms other than that controlled by pKs of chloroquine and the pH difference between the parasite food vacuole and the external medium have been postulated (Krogstad & Schlesinger, 1986; Sanchez, Wunsch, & Lanzer, 1997).

The high accumulation of chloroquine results in raising the pH of the food vacuole above the level of the acid proteases' optimal activity. This could inhibit the proteolysis of hemoglobin, the main source of amino acids for normal growth and replication. Chloroquine's effect on raising the pH of the food vacuoles was postulated to be its primary action as an antimalarial agent. However, direct correlation between changes in the pH of food vacuoles and their digestive function and the antimalarial action of chloroquine has not been established. It appears that the change in pH of food vacuoles does not completely explain its antimalarial effect.

Heme Polymerization: A Target for Chloroquine Action

Although there is a general agreement that chloroquine-sensitive plasmodia in mammalian erythrocytes can maintain high concentrations of the antimalarial agent, the molecular mechanism of the selective toxic action of chloroquine on malaria parasites has not been fully resolved. Currently, the most plausible mechanism is thought to be related to the functioning of the food vacuole in the digestion of hemoglobin by intraerythrocytic stages of the malaria parasite. Hemoglobin is broken down by proteolysis to heme and globin in the food vacuoles of the plasmodia.

The malaria parasite does not have an enzyme system to open the porphyrin ring of heme and thereby detoxify it. The concentration of heme in the malaria food vacuoles has been estimated to be as high as 0.4 M. In the absence of chloroquine, heme is detoxified by its polymerization to an inert crystalline material called hemozoin or malaria pigment. Chloroquine has a great tendency to bind to heme, thereby inhibiting the process of heme polymerization. Heme produced from the digested hemoglobin, if not polymerized, is oxidized from the ferrous (+2) to the ferric (+3) state giving rise to hemin (ferriprotoporphyrin IX). The released electrons combine with molecular oxygen to produce reactive oxygen species (superoxide anions and hydroxyl radicals) that have a lytic effect on the parasite and the erythrocyte owing to oxidative stress. The toxic effect of hemin on these cells has been ascribed to its interaction with lipid bilayers. The studies by Schmitt related this effect to an alteration in the membrane bilayer permeability, which results in structural disorder of membrane function and lipid peroxidation (Schmitt, Frezzatti, & Schreier, 1993). Both effects can lead to increased permeability of the membrane and lysis of the erythrocytes and the parasite.

The process of polymerization of heme to hemozoin is, therefore, an important process for the survival of the parasite in the host. Electron microscopy of hemozoin shows that the polymer has an orderly fibril organization, which indicates hydrogen bonding between straight chains (Bohle et al., 1994). These polymeric forms of heme are released as harmless metabolic by-products of the parasite. Sullivan et al. (1996) have recently examined the mechanism of association of chloroquine with hemozoin using electron microscope autoradiography and subcellular fractionation. Their data indicate that chloroquine associates with heme, forming a chlorquine–heme complex that caps the hemozoin and blocks further polymerization of heme. The outcome is an accumulation of unpolymerized heme that is toxic to the parasite. The main reason for the selective effect against malaria is that chloroquine accumulates in the *Plasmodium* digestive vacuoles in almost millimolar concentrations whereas nonparasitized erythrocytes and other host cells do not hyperconcentrate the quinoline antimalarial agents.

Originally, it was thought that the process of heme polymerization is catalyzed by a special heme polymerase in the parasite (Slater & Cerami, 1992). This has been refuted by Dorn et al. (1995), who reported that the polymerization is an autocatalytic chemical process, enhanced by the acid pH of the parasite food vacuole and by certain phospholipids. It is also thought that histidine-rich proteins may participate in initiating the polymerization reaction. Although a histidine-rich protein II was shown to bind to heme and to initiate its polymerization, evidence for its obligatory participation in this process has not yet been demonstrated (Sullivan et al., 1996).

The mechanism of antimalarial action of quinine, mefloquine, and amodiaqine has not been studied as extensively as that of chloroquine. However, these agents appear to have the same effects as chloroquine. These antimalarial agents are weak bases and have been reported to accumulate inside the acidic food vacuole of the parasites, where they cause alkalinization (although to a lesser degree than chloroquine) and inhibit heme polymerization (Chou, Chevli, & Fitch, 1980). The great interest in this research problem highlights the importance of the process of detoxification of heme by polymerization and draws our attention to an important target for selection of new antimalarial drugs.

The Proteases of Hemoglobin Digestion

Malaria parasites depend on the host's hemoglobin as the prime source for amino acids. As soon as the merozoites enter the host's erythrocytes where

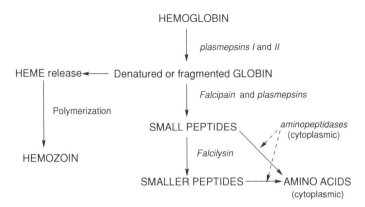

Figure 4.2. Schematic pathways of hemoglobin degradation in *P. falciparum*. Adapted from (Francis, Sullivan, & Goldberg, 1997) and (Eggleson, Duffin, & Goldberg, 1999).

there is a large amount of hemoglobin they start to change into ring forms and develop into trophozoites. Hemoglobin digestion in the food vacuoles occurs largely in trophozoites and early schizonts, the most metabolically active intraerythrocytic stages. The trophozoites ingest hemoglobin using a cellular structure called a cytostome, which is an invagination that extends across the parasite plasma membrane and the parasitophorous vacuole membrane that is derived from the erythrocyte during the invasion (Trager, 1986; Francis, Banerjee, & Goldberg, 1997). Initially many proteolytic enzymes were implicated in the process of hemoglobin digestion. It was not possible to distinguish those enzymes actually involved in hemoglobin digestion from those that participate in other digestive processes in the parasite. To eliminate proteolytic enzymes that are not involved in hemoglobin digestion it was essential first to isolate and purify the digestive vacuoles of the parasites. This was done by Goldberg and his associates for *P. falciparum* (Goldberg *et al.*, 1990). The process of hemoglobin ingestion by the trophozoite, its transport to the food vacuole of the parasite, and its digestion is an ordered one and is still under investigation. For review see Francis, Sullivan, & Goldberg (1997). Figure 4.2 shows the digestive process. First, the cytostomes ingest the host's hemoglobin. They then act as transport vesicles to carry it into the food vacuole. There plasmepsin I (formerly known as hemoglobinase I), which is an integral membrane protein, is cleaved to its soluble active form and starts the digestion of hemoglobin by separating the heme from the globin. As noted above heme is released and is sequestered in crystalline form as hemozoin. Another aspartic protease that can cleave native hemoglobin is plasmepsin II. Whereas plasmepsin I is present in the initial stage of hemoglobin digestion (ring stage), both plasmepsin I and II are required to complete the digestion of a massive amount of globin. The optimal activity of the two aspartic proteases takes place at pH 5.5–5.0, the physiological pH of the food vacuole. The outcome of this proteolytic digestion is the formation of large peptides from globin. A third enzyme, a cysteine protease known as falcipain, then digests the large peptide fragments into small peptides. This proteolytic enzyme is found in the trophozoite stage but not in the earlier ring stage of the parasites. Falcipain cannot cleave intact hemoglobin but can only degrade large globin peptides into smaller peptides. After transport of these peptides to the cytoplasm of the malarial parasite final digestion to amino acids is catalyzed by exopeptidases, which are optimally active at neutral pH.

Recently, Goldberg's laboratory reported that, in addition to the three proteases found in the malaria food vacuole, there is a fourth, a novel metallopeptidase, named falcilysin (Eggleson, Duffin, & Goldberg, 1999). This enzyme has been purified, cloned, sequenced, and characterized. It is an oligoendopeptidase that cleaves sites along the hemoglobin α and β chains. It cleaves peptides of up to twenty amino acids, preferentially those that are eleven to fifteen residues in length. It cannot cleave hemoglobin or globin. Therefore, it functions downstream from the other hemoglobin-digesting enzymes in the food vacuole. Falcilysin is inhibited by alkylating agents but not by the aminopeptidase inhibitors amastatin and bestatin. The importance of this finding is that falcilycin could be a new target for the selection of novel antimalarial agents. The final step in the hemoglobin proteolysis reactions, digestion of the small peptides to amino acids, occurs in the cytoplasm, which has a pH close to neutral. It is thought that proteolysis of these peptides is catalyzed by peptidases that are present in the parasite cytoplasm.

Malarial Hemoglobin Digestion as a Target. The dependence of malarial parasites on hemoglobin degradation as the main source of amino acids essential for their survival and multiplication drew attention to the different proteases involved and to the possibility of using them as antimalarial targets. None of the proteases is currently targeted by any known antimalarial drug. The World Health Organization has acknowledged this group of digestive proteases as appropriate targets for screening new antimalarial agents (WHO, 1996). Only during the current decade has enough information become available about the molecular components involved in hemoglobin digestion. Sophisticated investigations on different proteases have been followed by information on prospective inhibitors that may give us good leads to identifying new antimalarial agents.

Inhibitors of Hemoglobin Digestion. Cleavage of native hemoglobin in the food vacuole is the initial obligatory step for its digestion by the parasite. For

SC-50083 from Francis *et al.* (1994)

this reason it could be a good selective target for identifying antimalarial agents. Without the cleaved hemoglobin the other proteolytic enzymes will have no substrates to hydrolyze. After screening different peptidomimetic agents against plasmepsin I, Francis *et al.* (1994) found that the most potent inhibitor was SC-50083. The molecule is uncharged and has nonhydrolyzable amide bonds. This inhibitor has an IC_{50} of $(5-6) \times 10^{-7}$ M against plasmepsin I. It also inhibits plasmepsin II, the second aspartic protease, but only to a lesser extent. Since the peptide inhibits the initial reaction of hemoglobin digestion, all succeeding reactions of vacuolar hemoglobin digestion are impaired. This inhibitor kills 98% of malaria parasites cultured *in vitro* at a concentration of $10\,\mu$M when added just before the beginning of the trophozoite (ring stage) cycle, the stage of the parasite when hemoglobin digestion is carried out. One interesting aspect of this work was that the inhibitor was equally active against chloroquine-sensitive and chloroquine-resistant strains. This is not surprising because chloroquine acts against the parasite by inhibiting heme polymerization at a site that is different from where SC-50083 acts. One drawback of this inhibitor is that it showed some inhibition of the mammalian aspartic proteases, rennin and cathepsin, an indication that the inhibitor is not selective enough.

Research on proteases used for hemoglobin digestion has been advanced by the cloning of both plasmepsin I and II. The enzymes have sequence homology to mammalian aspartic proteases. For screening drugs plasmepsin II has been used as a primary target because sufficient quantities of this purified recombinant protein are available (Silva *et al.*, 1996). Pepstatin A is a known pentapeptide inhibitor of aspartic proteases and was also shown to inhibit hemoglobin degradation by extracts of malaria digestive vacuole. The crystal structure of plasmepsin II is also now available. A complex of pepstatin A and plasmepsin II was used to obtain crystallographic information about the structural framework of the enzyme–inhibitor interaction. Possible plasmepsin II inhibitors were identified on the basis of that structure. Based on this information a series of relatively low molecular weight plasmepsin inhibitors were screened and their effects on the growth of *P. falciparum* were examined. Compound 7 showed high potency as an inhibitor of plasmepsin II and was the best inhibitor of *P. falciparum* growth *in vitro*. It also appears to have high selectivity since it had the highest K_i against mammalian cathepsin.

Compound 7 from Silva *et al.* (1996)

The search for inhibitors of hemoglobin degradation in malaria parasites has recently been enhanced by the use of combinatorial libraries. To screen for plasmepsin inhibitors combinatorial libraries containing a statine core (transition state mimetic for aspartyl proteases) with a variety of different amino acid substitutions were synthesized. Varying 31 amino acids of the statine core structure provided great diversity of structure among the 13,020 member library (Carroll *et al.*, 1998b). The library was screened with both malarial plasmepsin II and mammalian cathepsin D to find the most potent and selective inhibitors. In addition, structure–activity relationships provided new information for finding inhibitors of plasmepsin II. Some of the compounds were shown to inhibit parasite growth in cell culture. Another combinatorial library was tested and this led to the identification of selective inhibitors for both malarial and mammalian proteases (Carroll *et al.*, 1998a).

Other researchers have been searching for low molecular weight nonpeptide protease inhibitors. These inhibitors would have an advantage over peptides because they are more likely to be rapidly absorbed by and have greater stability in the host tissues. Screening for this type of inhibitor has also been advanced by combinatorial chemistry and structure-based design. Inhibitors with molecular weights between 594 and 650 Da and with K_is of 2–10 nM and high selectivity as inhibitors of plasmepsin II have been identified (Haque *et al.*, 1999).

The cysteine protease (falcipain) from *P. falciparum* is another prospective target that has been investigated. This enzyme belongs to the papain cysteine protease family and is characterized by its acidic pH optimum and its specific substrates. Several fluoromethyl ketones that are known to inhibit other cysteine proteinases were tested as inhibitors of falcipain. Their effects on hemoglobin degradation and on the viability of *in vitro* cultures of *P. falciparum* were also tested (Rosenthal *et al.*, 1991). Among the fluoromethyl derivatives examined, benzyloxycarbonyl-Phe-Arg-CH$_2$F was the most potent ($K_i = 5$ pM). This inhibitor also blocked hemoglobin degradation ($K_{50\%} = 10$ pM) and killed the parasites *in vitro* (LD$_{50} = 6.4$ pM). As a proof of selectivity, this inhibitor was examined against four different human cell lines at concentrations up to 100 μM and showed no toxicity to the cells. The fluoromethyl ketone was reported to be effective *in vivo*; after a 4-day regimen it cured 80% of *Plasmodium vinckei–* infected mice (Rosenthal, Lee, & Smith, 1993). Correlation between the sensitivity of falcipain to inhibition by this antiprotease agent and its reported *in vivo* effect against murine infection indicates that falcipain is another prospective target to screen for selective antimalarial agents.

The optimal procedure to obtain tertiary structure of a protein target is X-ray crystallography. In many cases (e.g., the cysteine protease falcipain) it is not possible to obtain the pure protein in sufficient quantity to make the necessary crystals. One approach is to computer model the tertiary structure of the target on the basis of its sequence and the crystal structures of related enzymes. A successful collaborative effort between computational modelers and molecular pharmacologists identified low molecular weight nonpeptide inhibitors of falcipain of *P. falciparum* trophozoites. The 3-D structure was modeled on the basis of the predicted parasite protease sequence and the structural alignments of the proteolytic enzymes papain and actinidin (Ring *et al.*, 1993). The protease model structures were used to model interactions of ligand binding and to identify prospective inhibitors. One of these nonpeptide inhibitors, oxalic bis[(2-hydroxy-1-naphthylmethylene)hydrazide], was used as a lead compound. This compound inhibited the trophozoite protease with an IC$_{50}$ of 6 μM. The metabolism of parasites cultured *in vitro* was inhibited at the same concentration.

Oxalic bis[(2-hydroxy-1-naphthylmethylene)hydrazide]
from Ring *et al.* (1993)

Although the lead compound is not yet used as an antimalarial, it is an initial step in further drug development. The fact that this compound inhibits both enzymes and intact parasites at micromolar concentrations was an impetus for further synthetic and screening efforts, especially of chalcones (condensations of aldol, acetophenone, and benzaldehyde) and phenothiazines (Rosenthal, 1999).

Resistance to Chloroquine

Emergence of strains of *P. falciparum* that are resistant to chloroqine has been observed since shortly after its introduction as an important antimalarial agent. During the past twenty years resistance to chloroquine has become widespread in all areas where *P. falciparum* is endemic. In these places chloroquine and related quinoline derivatives are now essentially useless. In addition, chloroquine-resistant strains of *P. vivax* are now being reported in both Southeast Asia and South America. Where malaria has become resistant to chloroquine alternative antimalarial agents such as mephloquine, halofantrine, artemisinin derivatives, and the combination of pyrimethamine and sulfadoxine (Fansidar) are used. There is genetic diversity among wild strains of *P. falciparum* that infect humans in a given area. Thus there are differences in their sensitivity to chloroquine and other antimalarial agents. Resistance appears to emanate from spontaneous chromosomal point mutations. Resistant mutants are able to survive in the presence of antimalarial drugs. Although details of the mechanisms of resistance to chloroquine have yet to be established, it is well recognized that erythrocytes harboring resistant strains of malaria are able to concentrate less chloroquine than sensitive strains. There is no significant difference in the values for vacuole pH between chloroquine-resistant and chloroquine-sensitive strains of *P. falciparum* to account for the decrease in chloroquine concentration (Krogstad & Schlesinger, 1986). The mechanism of such a decrease in accumulation of chloroquine by resistant stains has been the subject of many investigations during the past decade with no conclusive results. Indirect experimental evidence came from the finding that the chloroquine-resistant phenotype is characterized by an increase in drug efflux rate compared to sensitive parasites. Resistant parasites pump chloroquine out 40–50 times faster than sensitive parasites.

Multidrug Resistance in P. falciparum. In human cancers resistance to chemotherapy by broad-spectrum anticancer agents is called multidrug resistance (MDR). This type of resistance has been reproduced experimentally in animal and human cells. One characteristic feature in these cells, as well as in human cancers, is the pattern of cross-resistance. The cells become resistant to a wide variety of cytotoxic agents that have completely different chemical structures and characteristics. The lower concentration of the anticancer drug in the cells could be caused either by a decrease in the influx of the drug or an increase in the efflux. Drugs to which these cells become resistant interact with a transporter that could be responsible for the resistance. MDR transporters, which are encoded by the *mdr* gene, are cell surface phospho-glycoproteins (P-glycoproteins). The *mdr* gene products (called either MDR_1 or PGY_1) have been proven to be responsible for both increased efflux and decreased influx of various chemotherapeutic agents, although it is generally understood that their function is to extrude cytotoxic drugs from MDR cells (Bray & Ward, 1998). There are certain agents that have been discovered to reverse the multidrug resistance phenotype by competing for the transport system responsible for resistance. These "reversing agents" or "chemosensitizers" include several calcium channel blockers such as verapamil and nifedipine (Gottesman & Pastan, 1993). The experience gained from studying MDR in cancer cells has been applied to chloroquine-resistant malaria and similarities have been noted.

In resistant *P. falciparum* verapamil and other calcium channel blockers slow the efflux of chloroquine, but they do not have that effect in sensitive strains (Martin, Oduola, & Milhous, 1987). A homolog of the mammalian *mdr* gene was isolated from *P. falciparum* and called *pfmdr1*. Sequence comparison of the mammalian and malarial genes shows very good homology, especially in the two regions that are the predicted ATP binding sites. A second *pfmdr2* gene appears to be quite different from its mammalian homolog and also different from *pfmdr1*. There is little evidence of a role for *pfmdr2* in chloroquine resistance (Bray & Ward, 1998). However, there is some evidence linking chloroquine resistance to a mutation of *pfmdr1* (Foote *et al.*, 1989). The gene is amplified in some chloroquine-resistant isolates of *P. falciparum* just as the *mdr* gene is amplified in MDR tumor cells. Foote *et al.* found that the complete sequences of *pfmdr1* from two chloroquine-sensitive isolates are identical. In five different chloroquine-resistant isolates, one to four key nucleotide differences resulted in amino acid substitutions. In a blind study they were able to correctly predict the chloroquine sensitivity or resistance of thirty-four out of thirty-six isolates on the basis of the amino acid substitutions they identified in the resistant strains.

In contrast to these findings Wellems *et al.* (1990) reported that in a genetic cross experiment they found no linkage between the chloroquine-resistant phenotype and *P. falciparum mdr* genes. The results suggest that resistance to chloroquine could be multigenic and that there is a determinant of resistance that does not involve MDR. It is obvious that more work is needed to understand the molecular mechanism and genetics behind the phenomenon of chloroquine resistance.

Resistance of P. falciparum *to Other Antimalarial Agents.* Some strains of malaria parasites are resistant to antimalarial agents of diverse chemical structures, but there is also significant variability as to the resistance or sensitivity of these strains to different quinoline derivatives (Bray & Ward, 1998). For example, two strains of *P. falciparum* that were selected in the laboratory for their high level of chloroquine resistance became more susceptible to mefloquine, a closely related quinoline antimalarial agent. Moreover, selection for resistance to halofantrine, a phenanthrene methanol derivative, which has a structure related to mefloquine, produced a mutant that is more sensitive to chloroquine than the original cells. In addition to these laboratory findings, isolates from the field can show an inverse relationship between chloroquine and mefloquine resistance. Clinical field reports show reduced chloroquine resistance in geographical areas where resistance to mefloquine is increasing. The data from both laboratory experiments and field reports indicate that cross-resistance of *P. falciparum* to antimalarial agents does not necessarily follow the same pattern seen in cross-resistance of mammalian cells to anticancer agents in which MDR genes play an important role. For interpretation of these differences between cross-resistance among malaria strains and cancer cells, an important factor that should be taken into consideration is that interference with hemozoin formation within the digestive vacuole of the malaria parasite requires that chloroquine or other aminoquinolines must be transported across several membrane bilayers within the mammalian host and into the malaria parasite. These include the erythrocyte plasma membrane, the parasitophorous vacuole membrane, the parasite plasma membrane, and the membrane of the digestive vacuole (Goldberg *et al.*, 1997). The contributions of all these structures to parasite sensitivity or resistance have yet to be fully understood. Mutations in any of these structures added to a change in the efflux of the drug by the MDR mechanism could contribute to changes in the sensitivity of the parasite to different antimalarial agents. The case of cross-resistance to antimalarial agents appears to be more complex than that seen in cancer cells.

Chloroquine Uptake by Resistant Plasmodium falciparum. A decrease in the uptake of chloroquine by resistant strains of *P. falciparum* could very well account for the loss of the sensitivity of the parasites to some antimalarial agents. Although this has not been proven beyond any doubt, recent investigations by Sanchez *et al.* (1997) shed some light on the possible role of chloroquine uptake on resistance to this antimalarial drug. This group reported that accumulation of chloroquine by *P. falciparum*–infected erythrocytes occurs through a carrier-mediated transport mechanism. Uptake and accumulation in the parasites (which includes binding to specific sites) were shown to have saturable kinetics, temperature dependence, and sensitivity to specific inhibitors. The reported features of the specific binding sites indicate that the carrier involved in chloroquine uptake is a plasmodial Na^+/H^+ exchanger. The kinetics of chloroquine uptake by chloroquine-sensitive and resistant strains were interpreted by the Michaelis–Menten equation to be of first order. Significant differences between the kinetic parameters for the two malaria strains were found. Maximal apparent transport rates for sensitive parasites were shown to be significantly higher than those for resistant strains. In addition, the Michaelis–Menten constant (K_m) is one order of magnitude lower than that of the resistant strains. These kinetic differences partially explain the reduced accumulation by chloroquine-resistant strains of malaria. More support for this conclusion comes from analysis of genetic crosses between sensitive and resistant strains. The chloroquine-resistant phenotype was shown to link with changes in chloroquine uptake kinetics. Indications that chloroquine uptake is mediated by a Na^+/H^+ exchanger came from studies on the effect of specific inhibitors of chloroquine uptake. For example, amiloride and its derivatives, such as 5-(N,N-dimethyl) amiloride and 5-(N-ethyl-N-isopropyl) amiloride, were shown to competitively inhibit chloroquine uptake. These same compounds are known also to be inhibitors of a Na^+/H^+ exchanger protein and compete for the same binding sites. More biochemical and physiological evidence reported by the Lanzer group (Wunsch *et al.*, 1998) suggests that chloroquine activates the plasmodial Na^+/H^+ exchanger in chloroquine-sensitive *P. falciparum*. This results in a transitory phase of rapid Na^+/H^+ exchange during which chloroquine is taken in and transported. However, the inability of chloroquine to stimulate its own uptake by the Na^+/H^+ exchanger may be the limiting step in the generation of a chloroquine-resistant phenotype. Such a modification of the exchanger in resistant strains could result from mutation within the protein itself, or in other accessory regulatory enzymes or binding proteins involved in modulating the Na^+/H^+ exchange system. More information regarding the structure and function of the exchanger system is needed before we can fully understand the molecular mechanisms of malaria resistance to chloroquine and related antimalarial agents. Investigations on the molecular basis of drug uptake by the malaria parasites may contribute to a better design of antimalrial agents, particularly drugs to which the parasites will not become resistant.

Antimalarials Used against Chloroquine-Resistant Malaria

Halofantrine and Mefloquine

Because malaria strains in diverse places became resistant to antimalarial agents there have been many attempts to develop new drugs that act on new targets. Halofantrine, which was originally discovered in the 1940s, was introduced by the Walter Reed Army Institute as an antimalarial agent against chloroquine-resistant malaria in the 1980s (White, 1996). It is a phenanthrene methanol derivative that was found to be effective as a blood schizonticide against chloroquine-resistant *P. falciparum* and *P. vivax* strains. It is of limited use in patients with arterioventricular conduction and ventricular repolarization problems. Halofantrine is particularly useful when mefloquine has proven ineffective.

HALOFANTRINE   MEFLOQUINE

Mefloquine is another contribution by the Walter Reed research program. It is a 4-quinoline-methanol derivative structurally related to quinine. Although mefloquine is effective against the erythrocytic stages of all malaria strains its use is usually restricted to erythrocytic stages of chloroquine-resistant *P. falciparum*. Unfortunately, there are reports of strains of *P. falciparum* developing resistance to mefloquine among populations that have been exposed to the drug. Mefloquine is thought to act by a mechanism similar to that of chloroquine, inhibiting heme polymerization during hemoglobin digestion in the parasite food vacuole. Clinical resistance to halofantrine and mefloquine appears to be spreading.

Artemisinin (Qinghaosu)

Artemisinin is the active principle obtained from a shrub called *Artemesia annua* (sweet wormwood) or qing-hao (pronounced ching-how) that has been used in China for twenty centuries as a decoction for the treatment of fever and chills.

ARTEMISININ

In 1967 the government of the People's Republic of China established a research program to study the antimalarial action of *Artemesia* as well as other medicinal plants. Purification and crystallization of the antimalarial active principle led to the isolation of Qinghaosu, which was renamed "artemisinin." The history and chemistry of this compound and related derivatives have been reviewed by Klayman (1985). Artemisinin (a sesquiterpene) is one of the few naturally occurring compounds that contains a peroxide. The marked difference between the structure of artemisinin and the other known antimalarial agents is probably why there are no reports of cross-resistance between the two categories of drugs. There is also no reported resistance to artemisinin in human malaria infections.

The mechanism of action of artemisinin has recently been reviewed (Meshnick *et al.*, 1996). Artemisinin derivatives show high selective toxicity against the malaria parasite. Whereas micromolar concentrations of artemisinin are needed for toxicity to mammalian cells, nanomolar concentrations are toxic to the malaria parasites. One factor involved in the selective toxicity is that the erythrocytes infected with *P. falciparum* concentrate artemisinin as well as its derivatives to concentrations 100-fold higher than do uninfected erythrocytes. Because of their hydrophobic properties, artemisinin and its congeners tend to bind to different membranes of the parasite, including those of the digestive vacuole and the mitochondria. There is strong evidence to indicate that the lethal effect of artemisinin is due to its production of free radicals, a process that is catalyzed by heme and iron. Artemisinin analogs lacking the endoperoxide bridge (which is known to be the source of oxygen free radicals) lack antimalarial effects (Klayman, 1985). The evidence is strong that artemisinin and its analogs are activated by heme or molecular iron, leading to formation of free radicals or electrophilic (alkylating) intermediates. These reactive molecular species bind to and damage specific malaria membrane-associated proteins, which have not yet been identified.

Quinine

Quinine, like artemisinin, was introduced to Western medicine long before malaria parasites were identified. (For its mechanism of action see the beginning of this chapter.) Currently, quinine is mostly used against chloroquine-resistant strains of *P. falciparum*. It acts primarily as a blood schizonticide. Although there are some reports of quinine resistance in different parts of the world, particularly Southeast Asia, it is still effective against chloroquine-resistant *P. falciparum* in most other countries.

Primaquine

During his notable research that helped to establish the field of chemotherapy, Paul Ehrlich discovered that methylene blue is able to stain malaria cells in the blood without staining cells in other tissues. His attention was drawn to the possibility of exploiting this observation in the search for selective antimalarial agents. Here was a chemical that binds to the parasite, and presumably can damage it, but was not binding to the host cells. Guttman and Ehrlich found that methylene blue has a slight antimalarial effect in laboratory animals (Guttmann & Ehrlich, 1891). More important, this minor practical finding of the antimalarial effect of methylene blue emphasized the importance of the scientific rational method in drug discovery. German chemists in the dye industry were stimulated to prepare analogs that would be more effective and less toxic. This was the process that led to the synthesis of 8-aminoquinolines, which were then tested against avian malaria infections. Of this series, pamaquine was the first to be introduced as an antimalarial agent. More research in the United States during World War II resulted in the discovery of primaquine, which was successfully tried in the field and is still in use today.

Primaquine is a special antimalarial agent because it has no significant effect against the erythrocytic stages of malaria, but it kills the dormant hypnozoites lying within the hepatocytes (see the life cycle in Fig. 4.1). These stages may

$$CH_3$$
$$HNCH (CH_2)_3NH_2$$

PRIMAQUINE

persist even after other stages have been eliminated from the blood. It is the only antimalarial agent known, so far, that can destroy the hepatic stages and the latent tissue forms of the malaria parasites. For a complete cure of *P. vivax* and *P. ovale*, primaquine must be given in combination with antimalarial agents such as chloroquine that will eliminate the erythrocytic stages. Primaquine is also effective against gametocytes of all four species of plasmodia. This is useful in stopping the life cycle at that stage and preventing transmission of infection to the Anopheles mosquitoes.

In spite of the fact that 8-aminoquinolines are some of the early antimalarial agents to be discovered we have no definite information about their mechanism of action because attempts to culture the liver stages of *P. vivax* or *P. ovale in vitro* have not been successful. Fortunately, the general use of primaquine as an antimalarial agent has not resulted in any noticeable problems of resistance.

Antifolates and Their Targets

Folate cofactors are essential for the transfer of 1-carbon units in the metabolic processes of amino acid synthesis as well as in purine and pyrimidine synthesis. The erythrocytic stages constitute a major part of the asexual life cycle of the malaria parasites in the host. There is rapid growth with repeated divisions of the nucleus accompanied by a high rate of DNA and protein syntheses. This is accompanied by a great need for folate synthesis. *P. falciparum* can synthesize tetrahydrofolate from GTP, *p*-aminobenzoic acid (PABA), and l-glutamate (Krungkrai, Webster, & Yuthavong, 1989). There is evidence to indicate that the malaria parasites utilize both the *de novo* and the salvage biosynthesis of folate. Many of the enzymes that participate in the *de novo* biosynthetic pathway and the salvage pathway have been identified (Fig. 4.3). It has also been demonstrated by Krungkrai that folate can be salvaged from exogenous folate products such as *p*-aminobenzoylglutamate or pterin-aldehyde when these are added to malaria cultures *in vitro*. These folate products can be metabolized by the appropriate enzymes in the *de novo* folate biosynthetic pathway.

Dihydropteroate Synthase

Attention was originally drawn to the importance of folate as a target following the discovery of sulfanilamide as an antibacterial agent. Subsequently, it was established that PABA is an essential metabolite for the synthesis of folate and that sulfonamides act as structural analogs of PABA to inhibit its utilization by the malaria parasite as well as by many bacterial species.

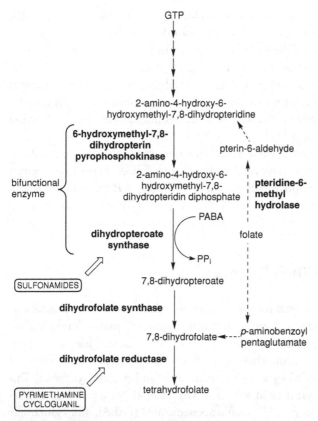

Figure 4.3. Schematic pathways of some of the reactions of *de novo* and salvage for the synthesis of tetrahydrofolate in *P. falciparum*. Solid arrows indicate the *de novo* pathway and dashed arrows indicate the salvage pathway. The sites of action of sulfonamides, pyrimethamine, and cycloguanil are shown. Adapted from Cowman (1988, pp. 317–330).

In papers that made history in the 1940s and 1950s D. D. Woods reported that sulfonamides owe their antibacterial action to the chemical similarity between their structure and the structure of PABA, an essential metabolite for the growth of sensitive bacteria. These experiments also showed the competitive kinetic relationship between the concentration of sulfonamide that inhibits bacterial growth and that of PABA needed by the microorganisms in the cultures. This work heralded a new field for using chemical analogs of essential metabolites as antibacterial or antiprotozoal agents. This approach is summarized in Woods's lecture at Oxford as a case history at the frontiers of biological discovery (Woods, 1962).

Sulfonamides, when given alone to birds infected with bird malaria, show a good antimalarial effect. In humans, however, sulfonamides do not give a complete cure unless they are combined with other antimalarial agents. Usually Fansidar (a combination of pyrimethamine and sulfadoxine) provides a synergistic therapeutic effect. Sulfadoxine is the most commonly used sulfonamide against malaria. The pharmacological reason for combining it with pyrimethamine is that the two compounds inhibit two adjacent reactions within the folic acid synthesis pathway. Sulfadoxine, as well as other sulfonamides, inhibits dihydropteroate synthase (DHPS), the enzyme that catalyzes the condensation of 6-hydroxymethyl-7,8-pterin pyrophosphate with PABA forming 7,8-dihydropteroate (see Fig. 4.3). Pyrimethamine and its congeners (e.g., trimethoprim) inhibit dihydrofolate reductase, which reduces dihydrofolate to tetrahydrofolate. These are the two major enzyme targets within the malaria folate biosynthetic pathway. Inhibition of the two enzymes acts synergistically, causing disruption of folate biosynthesis in the parasite. The outcome is a decrease in the pool of tetrahydrofolate cofactors, which results in a depletion of purine and pyrimidine synthesis and inhibition of DNA synthesis. Moreover, as a result of inhibition of transfer of 1-carbon units there is also inhibition of the synthesis of methionine and glycine. The combination of pyrimethamine and sulfadoxine was shown to inhibit both *de novo* and folate salvage pathways in the malaria parasites (Krungkrai *et al.*, 1989).

Although the sulfonamide targets in malaria and in bacteria have been known for several decades, more details of the nature of the malaria DHPS enzyme and the mechanism of action of the sulfonamides on the malaria parasite have been reported. Investigations by Zhang and Meshnick confirmed the role of sulfonamides against the growth of malaria parasites as well as their competitive inhibition of malarial DHPS with respect to PABA. Parasitized erythrocytes take up more sulfonamides than do uninfected erythrocytes and the uptake is proportional to the degree of parasitemia of the erythrocytes (Zhang & Meshnick, 1991). In addition to their direct inhibitory effect on DHPS sulfonamides can be used by the DHPS reaction as a replacement for PABA, acting as a "subversive" substrate and giving rise to toxic analogs of dihydropteroate, which would inhibit the enzyme and other subsequent enzyme reactions (Brown, 1962).

Dihydropteroate synthase is of special importance because it is an enzyme that is not found in the host but is found in the parasite and is crucial for its growth.

Recently, more information regarding the molecular nature of malarial DHPS has been uncovered as a result of cloning the enzyme and elucidating its primary structure and molecular organization (Triglia & Cowman, 1994). DHPS cloned from *P. falciparum* was found to be a bifunctional enzyme that encodes a protein of 83 kDa, which has two domains, one that is homologous with DHPS and the other that is homologous with dihydro-6-hydroxymethylpterin pyrophosphokinase (PPPK) (see Fig. 4.3). The enzyme from *P. falciparum* is abbreviated as PPPK-DHPS. Genetic data showed that there is no gene amplification of DHPS in sulfadoxine-resistant strains. The basis for clinical resistance to the sulfonamides is attributed to point mutations in five positions affecting four codons. DHPS from resistant strains of *P. falciparum* was expressed in *E. coli* and was shown to have a K_i for sulfadoxine that was 1000 times that of the sensitive strains. Amino acid differences in the DHPS enzyme of the sulfadoxine-resistant strains of *P. falciparum* appear to be central to the mechanism of resistance to sulfones and sulfonamides (Triglia *et al.*, 1997).

Dihydrofolate Reductase-Thymidylate Synthase (DHFR-TS)

The research of George H. Hitchings and Gertrude B. Elion in the 1950s and 1960s on selective inhibitors of dihydrofolate reductase and antimetabolites of reactants in the purine pathway led to the discovery of new drugs for the treatment of malaria (pyrimethamine and trimethoprim), herpes virus infection (acyclovir), and acute leukemia (6-mercaptopurine and thioguanine). Their fundamental contributions have been articulated in condensed versions of their Nobel lectures delivered in Stockholm in 1988 (Elion, 1989; Hitchings, 1989). Since their original discoveries, DHFR has become a recognized target for the selection of antiparasitic agents as well as antibacterial and anticancer agents.

Humans ingest unreduced or partially reduced folic acid in many fresh vegetables and fruits. Many of the reduced forms of folic acid are oxidized in the stomach to folic acid. Dihydrofolic acid reductase (DHFR) reduces folic acid and its partially reduced derivatives to tetrahydrofolate. Nonreduced folic acid and monohydrofolate are poorer substrates for the enzyme than dihydrofolate.

Many 2,4-diaminopyrimidines were synthesized in the early 1950s in the laboratory of Hitchings and Elion. Pyrimethamine and trimethoprim were found to be active against malaria. The antimicrobial effect of these agents can be antagonized competitively by folic acid, indicating that these compounds interfere with the synthesis of folic acid in both bacteria and malaria. These compounds exert their antimicrobial effect by selectively inhibiting DHFR. Some bacteria as well as malaria have to synthesize their own dihydrofolate from dihydropteroic acid. This is reduced to tetrahydrofolate by DHFR. Malaria and other protozoal parasites (*Crithidia*, *Leishmania*, *Trypanosoma*, *Plasmodia*, and *Eimeria*) have a bifunctional enzyme that catalyzes both reduction of dihydrofolate and the *de novo* synthesis of thymidylate (dTMP) from dUMP catalyzed by thymidylate synthase (TS). This single peptide has two separate active sites for the two reactions and is referred to as dihydrofolic acid reductase-thymidylate synthase:

$$\text{dihydrofolic acid} + \text{NADPH} + \text{H}^+ \xrightarrow{\text{DHFR}} \text{tetrahydrofolate} + \text{NADP}^+$$

$$\text{dUMP} + \text{tetrahydrofolate} \xrightarrow{\text{TS}} \text{dTMP} + \text{dihydrofolate}.$$

The teleology behind having the bifunctional enzyme in these parasites appears to be facilitation of channeling of tetrahydrofolate to the site of TS where it is needed on the same peptide. It is possible to inhibit one catalytic site without affecting the other site of the bifunctional enzyme. For example, methotrexate can inhibit only DHFR whereas fluorodeoxy UMP can inhibit TS irreversibly without affecting DHFR. In mammals DHFR and TS are two separable enzymes. The bifunctional enzyme, DHFR-TS, found in protozoa is a dimer with a molecular weight of 110–140 kDa. The DHFR domain is on the amino terminus and the TS domain is on the carboxy terminus. The two domains are separated by a junction peptide of varying size depending on the source (Garrett *et al.*, 1984). The unusual arrangement of these two enzymes made the bifunctional peptide a good target for antiprotozoal agents. The mechanism of antimalarial action of pyrimethamine, trimethoprim, chloroguanil, and related derivatives is based

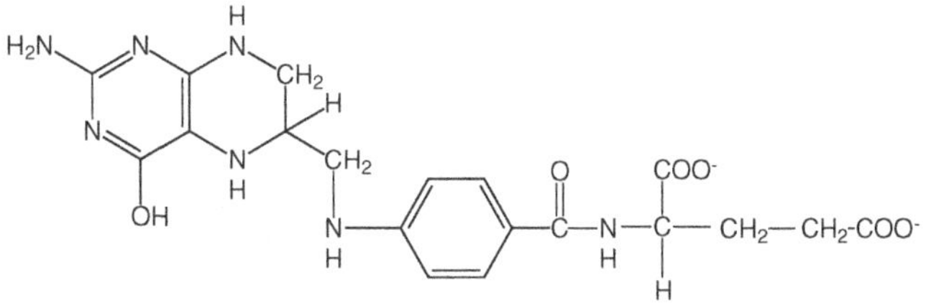

TETRAHYDROFOLATE

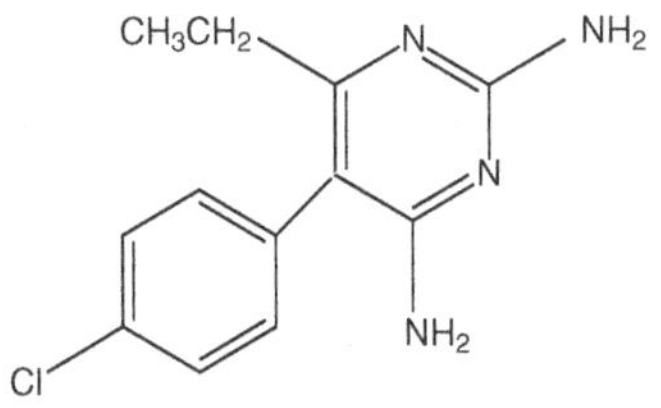

PYRIMETHAMINE

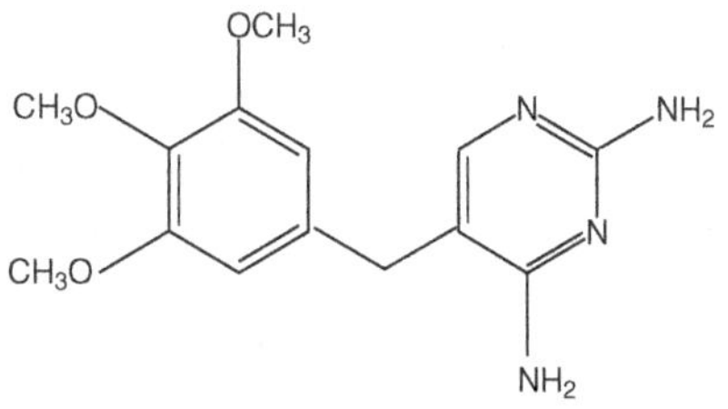

TRIMETHOPRIM

on selective binding and inhibition of the DHFR moiety of DHFR-TS. The lack of production of tetrahydrofolate inhibits the activity of TS, the second enzyme moiety of DHFR-TS. Such inhibition disrupts the DNA synthesis in the parasite by the depletion of dTMP. The antifolate effect of these compounds is enhanced when they are combined with a sulfonamide (e.g., sulfadoxine) or a sulfone.

Malaria Resistance to Pyrimethamine

Research on the genetics of pyrimethamine resistance showed a relationship between drug resistance and a point mutation in codon 108. Sequencing of the DHFR-TS gene from a large number of field isolates of *P. falciparum* identified a change of Ser-108 in pyrimethamine-sensitive isolates to Asn-108 in resistant isolates. Although most of the pyrimethamine-resistant isolates have Asn-108, some of the isolates that were highly resistant to pyrimethamine have additional mutations (Peterson, Walliker, & Wellems, 1988). The data indicate that pyrimethamine resistance arose as a single point mutation. Exposure of the parasite to high drug pressure selected for additional mutations that gave higher levels of drug resistance. The most significant feature of DHFR from resistant isolates is a marked increase in the inhibition constant (K_i) for pyrimethamine. Genetic confirmation of this mutation was a cross between pyrimethamine-resistant and pyrimethamine-sensitive parents. DHFR containing the Asn-108 was found to segregate with the determinant of the drug-resistant phenotype. Furthermore, transfection of the DHFR-TS gene containing Asn-108 and other alleles into *P. falciparum* was able to confer pyrimethamine resistance to the parasites (Wu, Kirkman, & Wellems, 1996).

The information about the molecular changes in the pyrimethamine resistance gene of *P. falciparum* has been exploited in screening strategies to identify new antifolate antimalarials (Brobey *et al.*, 1996). A synthetic gene that encodes for the truncated domain of DHFR of the bifunctional DHFR-TS was prepared using PCR amplification for the wild-type strains as well as for different resistant strains. The enzymes were expressed individually in *E. coli* and then were purified. The recombinant enzymes from wild-type strains and enzymes from antifolate-resistant strains were tested for their response to inhibition by several substituted pyrrolo[2,3-d] pyrimidines. The K_is for some 120 pyrimidine derivatives were tested on recombinant mutant and control enzymes and on bovine DHFR and compared with known DHFR inhibitors. The compounds that inhibited the DHFR from resistant strains were selected and tested for their inhibition of *in vitro* cultures of *P. falciparum*. This procedure has been successful in identifying a group of pyrrolo[2,3-d] pyrimidine antifolates that inhibited the growth of drug-resistant *P. falciparum*. The most selective derivatives of these compounds showed *in vivo* effects against malaria-infected mice.

Proteases and Merozoite Entrance and Exit from Erythrocytes

Proteases are involved in the process of merozoite invasion of erythrocytes and also in the exit of mature schizonts following rupture of the erythrocyte. Several enzymes have been implicated, but none has been as fully characterized as those involved in hemoglobin digestion. Reviews on the subject include those of Coombs & Mottram (1997) and Rosenthal (1999). The process of erythrocyte invasion starts by the merozoite positioning itself with its conical apical end in contact with the erythrocyte. A junction zone is then established between them where the contents of rhoptries are released from the merozoite. This is followed by invagination of the erythrocyte membrane, which forms the parasitophorous vacuole. The vacuole containing the merozoite is internalized into the erythrocyte.

Because the invasion process includes proteolysis and rearrangement of several membrane layers proteolytic enzymes and phospholipases are involved (Breton & Pereira da Silva, 1993). A *P. falciparum* serine protease (gp76) and its analog in a murine malaria *P. chabaudi* (gp68) are merozoite-specific serine proteases. The active enzymes are released from their membrane anchor, glycosyl-phosphatidylinositol, by a phosphatidylinositol-specific phospholipase C. Participation of the serine protease in invasion of merozoites has been confirmed using *P. chabaudi*, which is a good source for obtaining purified infectious merozoites. Experimental evidence indicates that the serine protease (gp68) is involved in erythrocyte invasion by specific degradation of an erythrocyte surface protein. Truncation of the surface protein appears to be necessary for successful entry of the merozoite into the erythrocyte. Preliminary identification of this protein indicates that it is the transmembrane anion transporter named erythrocyte band 3 (Breton *et al.*, 1992). Several peptides were synthesized and tried as substrates of the serine protease. The best substrate was found to be GlcA-Val-Leu-Gly-Lys-AEC. On the basis of the structure of this compound, peptidic inhibitors were synthesized by replacing the fluorescent amine present in the substrate by different alkylamines or amino alcohols. The most potent inhibitor is GlcA-Val-Leu-Gly-Lys-NHC_2H_5, which inhibits purified serine protease from *P. falciparum* with a K_i of 480 μM. The same compound was also potent against *in vitro* invasion of erythrocytes (IC_{50} = 900 μM) (Mayer *et al.*, 1991). The correlation between inhibition of serine proteases and inhibition of *in vitro* invasion of the red cells by the parasites is an indication that invasion is dependent on active serine proteases in the parasites. The IC_{50} value is probably not sufficiently low to warrant clinical trials of these inhibitors, but it is a good beginning for identifying a new target and deserves further investigation.

Plasminogen activators (PAs) are serine proteases that are associated with pathogenic bacteria and protozoal parasites. Urokinase (uPA) is a kind of plasminogen activator that was originally obtained from urine. It is a serine protease that cleaves plasminogen to active plasmin, which stimulates degradation of the erythrocyte cell surface to facilitate parasite invasion. Using both polyclonal and monoclonal antibodies raised against human urokinase it is possible to demonstrate that uPA binds to the surface of erythrocytes infected with mature forms of *P. falciparum*. When these cultures are depleted of urokinase the release of merozoites is inhibited and the erythrocytes become full of mature schizonts that are unable to break out of the erythrocytes. This inhibition can be reversed when urokinase is added to the medium (Roggwiller *et al.*, 1997). This recent information draws attention to a new plasminogen–plasmin enzyme system in erythrocytes and it could be utilized for identifying specific inhibitors.

There are several other proteases that have been implicated in the invasion process of merozoites and their release from the erythrocytes after schizont maturation. In-depth studies of these enzymes have been limited because of the meager supplies of material. From the limited information we have it seems clear that their properties are quite different from those of their mammalian counterparts and that they would probably be good targets for selective chemotherapeutic agents (Breton & Pereira da Silva, 1993).

Antisense Oligodeoxynucleotides

In keeping up with the current grand revolution in molecular biology, scientists who are interested in chemotherapy of cancer and bacterial or viral infections have been looking at antisense oligodeoxynucleotides (AS ODNs) as potential chemotherapeutic agents. The technology involves synthesizing modified oligodeoxynucleotides that would hybridize with the appropriate RNA target. For example, the internucleoside phosphate groups of the oligodeoxynucleotides contain either phosphodiester (PO) or phosphorothioates (PS) to provide stability against hydrolysis by host nucleases. Rapaport *et al.* (1992) have reported on the antimalarial effect of oligodeoxynucleotide containing phosphorothioates. The target used in these studies was the bifunctional enzyme DHFR-TS of *P. falciparum*. The synthetic oligodeoxynucleotides consisted of twenty-one nucleotides chosen to be complementary to selected sequences of the

P. falciparum DHFR-TS gene. The effects of these AS ODNs were tested using *in vitro* blood cultures infected with chloroquine-sensitive or chloroquine-resistant strains of *P. falciparum*. The first fortunate finding is that uninfected human erythrocytes present in the malaria cultures were impermeable to the AS ODNs being tested. It is fortunate because, usually, in human malaria, only 10–20% of the erythrocytes in the circulation are infected. Thus, in an *in vivo* setting the major population of the erythrocytes would not be exposed to possible toxic complications from the AS ODNs. The main finding of Rapaport and his colleagues was that the AS ODNs containing phosphorothioates that had a complementary sequence "antisense" directed against mRNA of the DHFR-TS gene were effective in inhibiting erythrocyte invasion and growth of both chloroquine-sensitive and chloroquine-resistant parasites in culture. The AS ODNs reduced parasitemia and hypoxanthine incorporation (a sign of growth) by *P. falciparum* in cultures. This inhibitory effect is due to the complementarity between the malaria mRNA for the DHFR-TS gene and the antisense oligodeoxynucleotide. Presumably, inhibition of translation by hybridization is accompanied by accelerated degradation of the target mRNA by activated ribonuclease H. Optimal selectivity of the AS ODNs is concentration dependent. At low concentrations (0.005 and 0.5 μM) the inhibition by the AS ODNs is directed specifically to the DHFR-TS gene, whereas at the higher concentrations (1 μM or higher) other genes were affected. The reason for the non-target-specific inhibition by these oligodeoxynucleotides has been postulated to be due to their charged polyanionic property (Barker *et al.*, 1996). Further experiments were carried out to investigate the relationship between the structure of the AS ODNs and their efficacy against malaria *in vitro*. It appears that limiting the number of phosphorothioates and inclusion of 2'-O-methyl groups may yield antisense compounds with higher activity as well as greater selectivity for the target (Barker *et al.*, 1998). One can foresee in the future using several antisense oligodeoxynucleotides, each of which could be directed to a sequence in a different gene. There could be a synergistic effect of simultaneously inhibiting both dihydropteroate synthetase and DHFR-TS.

The ingenious mechanisms that malaria parasites have adopted to evade the effects of different inhibitors of two crucial enzymes in the biosynthetic pathway of folate are most discouraging in the fight against this dreadful disease. However, the new knowledge gained, added to the current molecular biological technology recently acquired, gives some hope for a better future. Recent progress in current work on this aspect of the problem has made it possible to prepare unlimited amounts of recombinant enzyme preparations of DHPS and DHFR, two of the critical enzymes needed for these parasites to live in the host. This, considered with the new technology of crystallography and combinatorial chemistry, has opened up the possibility of selecting for other inhibitors of these targets.

REFERENCES

Barker, R. H., Jr., Metelev, V., Rapaport, E. & Zamecnik, P. (1996). Inhibition of *Plasmodium falciparum* malaria using antisense oligodeoxynucleotides. *Proc Natl Acad Sci USA*, *93*(1), 514–518.

Barker, R. H., Jr., Metelev, V., Coakley, A. & Zamecnik, P. (1998). *Plasmodium falciparum*: Effect of chemical structure on efficacy and specificity of antisense oligonucleotides against malaria in vitro. *Exp Parasitol, 88*(1), 51–59.

Bohle, S., Conklin, B., Cox, D., Madsen, S. & Paulson, S. (1994). Structural and spectroscopic studies of beta-hematin (the heme coordination polymer in malaria pigment). In P. Wisian-Neilson, H. Allcock, & K. Wynn (Eds.), *Inorganic and Organometallic Polymers II: Advanced Materials and Intermediates* (pp. 497–515). Washington, DC: Am. Chem. Soc.

Bray, P. G. & Ward, S. A. (1998). A comparison of the phenomenology and genetics of multidrug resistance in cancer cells and quinoline resistance in *Plasmodium falciparum*. *Pharmacol Ther, 77*(1), 1–28.

Breton, C. & Pereira da Silva, L. (1993). Malaria proteases and red blood cell invasion. *Parasitol Today, 9*, 92–96.

Breton, C. B., Blisnick, T., Jouin, H., Barale, J. C., Rabilloud, T., Langsley, G. & Pereira da Silva, L. H. (1992). *Plasmodium chabaudi* p68 serine protease activity required for merozoite entry into mouse erythrocytes. *Proc Natl Acad Sci USA, 89*(20), 9647–9651.

Brobey, R. K., Sano, G., Itoh, F., Aso, K., Kimura, M., Mitamura, T. & Horii, T. (1996). Recombinant *Plasmodium falciparum* dihydrofolate reductase-based in vitro screen for antifolate antimalarials. *Mol Biochem Parasitol, 81*(2), 225–237.

Brown, G. (1962). The biosynthesis of folic acid. *J Biol Chem, 237*, 536–540.

Carroll, C. D., Johnson, T. O., Tao, S., Lauri, G., Orlowski, M., Gluzman, I. Y., Goldberg, D. E. & Dolle, R. E. (1998a). Evaluation of a structure-based statine cyclic diamino amide encoded combinatorial library against plasmepsin II and cathepsin D. *Bioorg Med Chem Lett, 8*(22), 3203–3206.

Carroll, C. D., Patel, H., Johnson, T. O., Guo, T., Orlowski, M., He, Z. M., Cavallaro, C. L., Guo, J., Oksman, A., Gluzman, I. Y., Connelly, J., Chelsky, D., Goldberg, D. E. & Dolle, R. E. (1998b). Identification of potent inhibitors of *Plasmodium falciparum* plasmepsin II from an encoded statine combinatorial library. *Bioorg Med Chem Lett, 8*(17), 2315–2320.

Chou, A. C., Chevli, R. & Fitch, C. D. (1980). Ferriprotoporphyrin IX fulfills the criteria for identification as the chloroquine receptor of malaria parasites. *Biochemistry, 19*(8), 1543–1549.

Coombs, G. H. & Mottram, J. C. (1997). Parasite proteinases and amino acid metabolism: Possibilities for chemotherapeutic exploitation. *Parasitology, 114*(Suppl), S61–80.

Cowman, A. (1988). The molecular basis of resistance to the sulfones, sulfonamides, and dihydrofolate reductase inhibitors. In I. Sherman (Ed.), *Malaria – Parasite Biology, Pathogenesis and Protection* (pp. 317–330). Washington, DC: American Society for Microbiology Press.

Dorn, A., Stoffel, R., Matile, H., Bubendorf, A. & Ridley, R. G. (1995). Malarial haemozoin/beta-haematin supports haem polymerization in the absence of protein. *Nature, 374*(6519), 269–271.

Eggleson, K. K., Duffin, K. L. & Goldberg, D. E. (1999). Identification and characterization of falcilysin, a metallopeptidase involved in hemoglobin catabolism within the malaria parasite *Plasmodium falciparum*. *J Biol Chem*, *274*(45), 32411–32417.

Elion, G. (1989). The purine path to chemotherapy. *Science*, *244*, 41–47.

Foote, S. J., Thompson, J. K., Cowman, A. F. & Kemp, D. J. (1989). Amplification of the multidrug resistance gene in some chloroquine-resistant isolates of *P. falciparum*. *Cell*, *57*(6), 921–930.

Francis, S. E., Banerjee, R. & Goldberg, D. E. (1997). Biosynthesis and maturation of the malaria aspartic hemoglobinases plasmepsins I and II. *J Biol Chem*, *272*(23), 14961–14968.

Francis, S. E., Sullivan, D. J., Jr. & Goldberg, D. E. (1997). Hemoglobin metabolism in the malaria parasite *Plasmodium falciparum*. *Annu Rev Microbiol*, *51*, 97–123.

Francis, S. E., Gluzman, I. Y., Oksman, A., Knickerbocker, A., Mueller, R., Bryant, M. L., Sherman, D. R., Russell, D. G. & Goldberg, D. E. (1994). Molecular characterization and inhibition of a *Plasmodium falciparum* aspartic hemoglobinase. *EMBO J*, *13*(2), 306–317.

Garrett, C. E., Coderre, J. A., Meek, T. D., Garvey, E. P., Claman, D. M., Beverley, S. M. & Santi, D. V. (1984). A bifunctional thymidylate synthetase-dihydrofolate reductase in protozoa. *Mol Biochem Parasitol*, *11*, 257–265.

Goldberg, D. E., Slater, A. F., Cerami, A. & Henderson, G. B. (1990). Hemoglobin degradation in the malaria parasite *Plasmodium falciparum*: An ordered process in a unique organelle. *Proc Natl Acad Sci USA*, *87*(8), 2931–2935.

Goldberg, D. E., Sharma, V., Oksman, A., Gluzman, I. Y., Wellems, T. E. & Piwnica-Worms, D. (1997). Probing the chloroquine resistance locus of *Plasmodium falciparum* with a novel class of multidentate metal(III) coordination complexes. *J Biol Chem*, *272*(10), 6567–6572.

Gottesman, M. M. & Pastan, I. (1993). Biochemistry of multidrug resistance mediated by the multidrug transporter. *Annu Rev Biochem*, *62*, 385–427.

Guttmann, P. & Ehrlich, P. (1891). Uber die Wirking des Methyleneblau bei malaria. *Berliner Klin Woch*, *28*, 953.

Haque, T. S., Skillman, A. G., Lee, C. E., Habashita, H., Gluzman, I. Y., Ewing, T. J., Goldberg, D. E., Kuntz, I. D. & Ellman, J. A. (1999). Potent, low-molecular-weight non-peptide inhibitors of malarial aspartyl protease plasmepsin II. *J Med Chem*, *42*(8), 1428–1440.

Hitchings, G. (1989). Nobel lecture in physiology or medicine – Year 1988. *In Vitro Cell Dev Biol*, *25*, 303–309.

Klayman, D. L. (1985). Qinghaosu (artemisinin): An antimalarial drug from China. *Science*, *228*(4703), 1049–1055.

Krogstad, D. J. & Schlesinger, P. H. (1986). A perspective on antimalarial action: Effects of weak bases on *Plasmodium falciparum*. *Biochem Pharmacol*, *35*(4), 547–552.

Krogstad, D. J. & Schlesinger, P. H. (1987). Acid-vesicle function, intracellular pathogens, and the action of chloroquine against *Plasmodium falciparum*. *N Engl J Med*, *317*(9), 542–549.

Krungkrai, J., Webster, H. K. & Yuthavong, Y. (1989). De novo and salvage biosynthesis of pteroylpentaglutamates in the human malaria parasite, *Plasmodium falciparum*. *Mol Biochem Parasitol*, *32*(1), 25–37.

Macomber, P. B., O'Brien, R. L. & Hahn, F. E. (1966). Chloroquine: Physiological basis of drug resistance in *Plasmodium berghei*. *Science, 152*(727), 1374–1375.

Martin, S. K., Oduola, A. M. & Milhous, W. K. (1987). Reversal of chloroquine resistance in *Plasmodium falciparum* by verapamil. *Science, 235*(4791), 899–901.

Mayer, R., Picard, I., Lawton, P., Grellier, P., Barrault, C., Monsigny, M. & Schrevel, J. (1991). Peptide derivatives specific for a *Plasmodium falciparum* proteinase inhibit the human erythrocyte invasion by merozoites. *J Med Chem, 34*(10), 3029–3035.

Meshnick, S. R., Taylor, T. E. & Kamchonwongpaisan, S. (1996). Artemisinin and the antimalarial endoperoxides: From herbal remedy to targeted chemotherapy. *Microbiol Rev, 60*(2), 301–315.

Peterson, D. S., Walliker, D. & Wellems, T. E. (1988). Evidence that a point mutation in dihydrofolate reductase-thymidylate synthase confers resistance to pyrimethamine in falciparum malaria. *Proc Natl Acad Sci USA, 85*(23), 9114–9118.

Rapaport, E., Misiura, K., Agrawal, S. & Zamecnik, P. (1992). Antimalarial activities of oligodeoxynucleotide phosphorothioates in chloroquine-resistant *Plasmodium falciparum*. *Proc Natl Acad Sci USA, 89*(18), 8577–8580.

Ring, C. S., Sun, E., McKerrow, J. H., Lee, G. K., Rosenthal, P. J., Kuntz, I. D. & Cohen, F. E. (1993). Structure-based inhibitor design by using protein models for the development of antiparasitic agents. *Proc Natl Acad Sci USA, 90*(8), 3583–3587.

Roggwiller, E., Fricaud, A. C., Blisnick, T. & Braun-Breton, C. (1997). Host urokinase-type plasminogen activator participates in the release of malaria merozoites from infected erythrocytes. *Mol Biochem Parasitol, 86*(1), 49–59.

Rosenthal, P. J. (1999). Proteases of protozoan parasites. *Adv Parasitol, 43*, 105–158.

Rosenthal, P. J., Lee, G. K. & Smith, R. E. (1993). Inhibition of a *Plasmodium vinckei* cysteine proteinase cures murine malaria. *J Clin Invest, 91*(3), 1052–1056.

Rosenthal, P. J., Wollish, W. S., Palmer, J. T. & Rasnick, D. (1991). Antimalarial effects of peptide inhibitors of a *Plasmodium falciparum* cysteine proteinase. *J Clin Invest, 88*(5), 1467–1472.

Sanchez, C. P., Wunsch, S. & Lanzer, M. (1997). Identification of a chloroquine importer in *Plasmodium falciparum*. Differences in import kinetics are genetically linked with the chloroquine-resistant phenotype. *J Biol Chem, 272*(5), 2652–2658.

Schmitt, T. H., Frezzatti, W. A., Jr. & Schreier, S. (1993). Hemin-induced lipid membrane disorder and increased permeability: A molecular model for the mechanism of cell lysis. *Arch Biochem Biophys, 307*(1), 96–103.

Silva, A. M., Lee, A. Y., Gulnik, S. V., Maier, P., Collins, J., Bhat, T. N., Collins, P. J., Cachau, R. E., Luker, K. E., Gluzman, I. Y., Francis, S. E., Oksman, A., Goldberg, D. E. & Erickson, J. W. (1996). Structure and inhibition of plasmepsin II, a hemoglobin-degrading enzyme from *Plasmodium falciparum*. *Proc Natl Acad Sci USA, 93*(19), 10034–10039.

Slater, A. F. & Cerami, A. (1992). Inhibition by chloroquine of a novel haem polymerase enzyme activity in malaria trophozoites [see comments]. *Nature, 355*(6356), 167–169.

Sullivan, D. J., Jr., Gluzman, I. Y., Russell, D. G. & Goldberg, D. E. (1996). On the molecular mechanism of chloroquine's antimalarial action. *Proc Natl Acad Sci USA, 93*(21), 11865–11870.

Trager, W. (1986). *Living Together*. New York: Plenum Press.

Triglia, T. & Cowman, A. F. (1994). Primary structure and expression of the dihydropteroate synthetase gene of *Plasmodium falciparum. Proc Natl Acad Sci USA, 91*(15), 7149–7153.

Triglia, T., Menting, J. G., Wilson, C. & Cowman, A. F. (1997). Mutations in dihydropteroate synthase are responsible for sulfone and sulfonamide resistance in *Plasmodium falciparum. Proc Natl Acad Sci USA, 94*(25), 13944–13949.

Wellems, T. E., Panton, L. J., Gluzman, I. Y., do Rosario, V. E., Gwadz, R. W., Walker-Jonah, A. & Krogstad, D. J. (1990). Chloroquine resistance not linked to *mdr*-like genes in a *Plasmodium falciparum* cross [see comments]. *Nature, 345*(6272), 253–255.

White, N. J. (1996). The treatment of malaria [see comments]. *N Engl J Med, 335*(11), 800–806.

WHO. (1996). *P. falciparum Proteinases Inhibitor Development.* Geneva: World Health Organization.

Woods, D. (1962). The biochemical mode of action of the sulfonamide drugs. *J Gen Microbiol, 92*, 687–702.

Wu, Y., Kirkman, L. A. & Wellems, T. E. (1996). Transformation of *Plasmodium falciparum* malaria parasites by homologous integration of plasmids that confer resistance to pyrimethamine. *Proc Natl Acad Sci USA, 93*(3), 1130–1134.

Wunsch, S., Sanchez, C. P., Gekle, M., Grosse-Wortmann, L., Wiesner, J. & Lanzer, M. (1998). Differential stimulation of the Na^+/H^+ exchanger determines chloroquine uptake in *Plasmodium falciparum. J Cell Biol, 140*(2), 335–345.

Zhang, Y. & Meshnick, S. R. (1991). Inhibition of *Plasmodium falciparum* dihydropteroate synthetase and growth in vitro by sulfa drugs. *Antimicrob Agents Chemother, 35*(2), 267–271.

5

Antitrypanosomal and Antileishmanial Targets

The hemoflagellates of the genus *Trypanosoma* and of the genus *Leishmania* belong to the family of *Trypanosomatidae*. They are blood or tissue parasites that infect humans and other vertebrates. *Trypanosoma brucei gambiense* and *T. brucei rhodesiense* are the causative organisms of African trypanosomiasis (sleeping sickness), which is prevalent in central and eastern Africa. *T. brucei* can also infect all domestic animals. The vector of this disease is tsetse flies of the genus *Glossina*. Once the fly bites the mammalian host the metacyclic trypanosomes (mature stages in fly salivary glands) cause local skin "chancre" and then migrate to the lymphatic system of the host and eventually to the bloodstream. After approximately a month the parasites cross the choroid plexus into the brain and cerebrospinal fluid. The "sleeping" symptoms, such as physical depression, mental deterioration, and even coma, then appear in the host. The causative organism of American trypanosomiasis, also known as Chagas' disease, is *Trypanosoma cruzi*. It is present in South and Central America, especially in Brazil, Argentina, and Mexico. Vectors for transmission are members of *Triatominae*, insects found in Latin America. After having a blood meal from the human host the metacyclic forms of the trypanosome enter the gut of the insects and are eventually released in the insects' feces. When the feces are deposited on the skin of the host and rubbed they penetrate the skin and enter host cells where they are transformed to amastigotes, which cause chronic local inflammatory swelling. They multiply in the cells and transform to trypomastigotes that enter the host's blood and circulate. They may be ingested from the host during a blood meal by the vector insect.

In the case of infections by *Leishmania spp.* the parasite is transmitted by the bite of female sand flies of the subfamily *Phlebotominae*. Once the parasites are in the gut of the vector they are transformed to elongated flagellated promastigotes and quickly begin to multiply. In the host the parasites live intracellularly within the reticuloendothelial cells and are referred to as amastigotes. Most research on metabolism and the effects of drugs is done on promastigotes because these cells are easy to maintain in the laboratory. There are two main forms of leishmaniasis, the cutaneous and the visceral. The cutaneous form is caused by *Leishmania tropica*, *L. major*, and the New World *L. braziliensis*. The symptoms are characterized by the formation of cutaneous ulcers that are difficult to treat and leave ugly scars even after prolonged treatment. Visceral leishmaniasis, also known as Kala azar, is caused by *L. donovani* and is widespread in the tropics and subtropics. After entering the host's body infective promastigotes are taken up into macrophages. The infection eventually spreads to the local lymph glands, and reticuloendothelial cells in the spleen, liver, bone marrow, lymph glands, and other organs with serious systemic complications (Smyth, 1994).

Tripanothione and Tripanothione Reductase in Protozoa

All living organisms contain high levels of thiol-containing enzymes and other proteins. The integrity and function of these proteins depend on the presence of reduced glutathione, a tripeptide with a reduced sulfhydryl group. This is the most ubiquitous sulfhydryl buffer that maintains the cysteine residues of the sulfhydryl-containing proteins in the reduced state. To ensure that the SH groups in the cells are kept in the reduced form several hundred times more glutathione exists in the reduced form (GSH) than in the oxidized form (GSSG). Trypanosomatids differ from all other organisms in having a different peptide that performs the same function as GSH. In these organisms glutathione is replaced by trypanothione [N^1,N^8-bis(glutathionyl)] spermidine, a tripeptide composed of glutathione and spermidine. For a comprehensive review see Fairlamb & Cerami (1992). The oxidized and reduced states of glutathione and trypanothione are shown in Fig. 5.1. The figure shows the structural formulas of the human and parasite substrates of glutathione reductase (GR) and of trypanothione reductase (TR) before and after reduction by the appropriate enzyme.

Another important role played by GSH in the cell is that of detoxification of harmful oxidizing agents such as hydrogen peroxide and other organic peroxides. High intracellular levels of NADPH favor glutathione in the reduced form. In the parasite, trypanothione is present predominantly as the dithiol, dihydrotrypanothione. This reduction is catalyzed by the NADPH-dependent enzyme trypanothione reductase (TR), which is selective for trypanothione and cannot catalyze the reduction of glutathione disulfide (GSSG). In trypanosomes, trypanothione and its reductase provide the sulfhydryl buffering capacity to maintain the cysteine residues of the parasite proteins in the reduced forms.

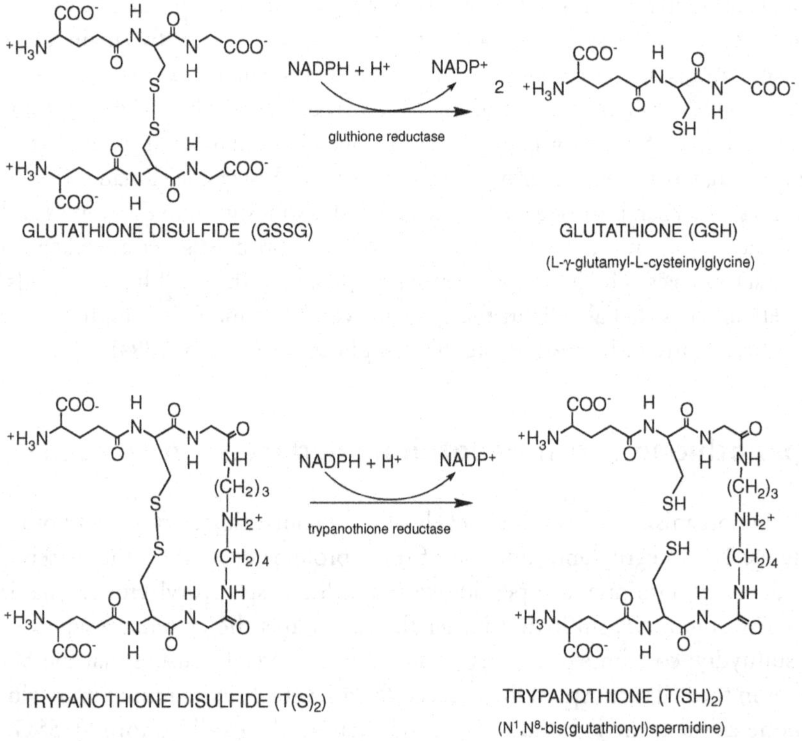

Figure 5.1. Structures of glutathione disulfide and trypanothione disulfide and the reactions catalyzed by human glutathione reductase and trypanosomatid trypanothione reductase according to Fairlamb & Cerami (1992).

The same system is also involved in defense against damage by oxidants, heavy metals, and xenobiotics.

An impressive demonstration of the role of TR in *in vitro* cultures as well as infectivity and virulence of the parasites was reported recently (Krieger *et al.*, 2000). In *T. brucei* cells levels of TR could be manipulated by molecular biological techniques. Trypanosomes having less than 10% of the wild-type TR activity were unable to grow in culture and showed hypersensitivity to H_2O_2. Furthermore, both the infectivity and the virulence of the parasites in mice were dependent on the level of TR activity. TR plays the same important redox defense role in *Leishmania*. Mutants of *L. donovani* with lower levels of TR activity showed attenuated infectivity and decreased capacity to survive within the macrophages (Dumas *et al.*, 1997). Because neither trypanothione nor its reductase are found in the mammalian host, the parasite sulfhydryl reducing system is considered to be a selective target for antiparasitic agents.

Trypanothione Reductase as a Target for Drug Design

Both mammalian GR and protozoal TR are homodimers of molecular mass of 100–110 kDa. The two enzymes share about 40% sequence identity including the residues necessary for enzyme catalysis. In spite of this close similarity between the two enzymes, each enzyme is specific for the cognate substrate. The difference in specificity of the two enzymes offers the opportunity for chemical design of inhibitors to the TR that do not affect GR. Characterization of the 3-D structure of the enzyme, particularly the active site, advanced the chances of designing selective inhibitors of the enzyme. *Crithidia fasciculata*, a protozoal parasite of insects, which is much easier to culture in the laboratory compared to pathogenic trypanosomes, has been used to isolate milligram amounts of pure TR required for crystallography. The suitability of this enzyme has been enhanced by the finding that the amino acid sequence of *C. fasciculata* TR shows 69% identity with that of *T. cruzi*. Crystallographic studies on TR of *C. fasciculata* showed similar topology to mammalian GR but also showed an enlarged active site and a number of amino acid differences that contribute to steric and electrostatic characteristics of these two enzymes (Hunter *et al.*, 1992). The enlarged active site allows it to accommodate the unique substrate trypanothione, but not glutathione. TR from *T. cruzi* was also crystallized and compared with mammalian glutathione reductase. The crystallographic information provides details of the mechanism of enzyme catalysis that could be used for development of transition-state analogs or inhibitors. Several structural characteristics differentiate the mammalian enzyme from that of the parasite. These differences are essential for the rational design and screening of trypanothione reductase inhibitors (Hunter *et al.*, 1992; Zhang *et al.*, 1993, 1996). Some selective inhibitors of trypanothione reductase from *Trypanosoma cruzi* have been rationally designed (Garforth *et al.*, 1997). Some of these tricyclic compounds containing the saturated dibenzazepine nucleus were found to have K_i values in the low micromolar range. This series was designed by using molecular graphic analysis of a 3-D homology model of the active site. The 3-D model has also been used in searching databases. Two cyclic-polyamine natural products were identified as prospective TR inhibitors. One was subsequently found to inhibit TR *in vitro* (Bond *et al.*, 1999).

Biosynthesis of Trypanothione

In African trypanosomes, *Leishmania spp.* and *C. fasciculata* spermidine is synthesized from ornithine and S-adenosylmethionine using the same pathway that mammals use to synthesize glutathione (Fairlamb & Cerami, 1992). The subsequent steps for the synthesis of trypanothione from glutathione and spermidine are not found in mammalian cells. The intermediate N^1, N^8-isomer glutathionyl spermidine is synthesized by enzymatic conjugation of glutathione and spermidine. Two distinct enzymes, glutathionyl spermidine synthase and trypanothione synthase, both requiring ATP, are involved in the conjugation step (Smith et al., 1992). Both enzymes are potential targets for antiprotozoal drugs (Fairlamb & Cerami, 1992).

Effect of Antitrypanosomal Agents on the Trypanothione System

There are two categories of drugs that are effective against trypanosomes owing to their action on the trypanothione system in the parasite. Organic arsenicals, which have been used as antitrypanosomal agents since Paul Ehrlich's time, are still in the medical and veterinary therapeutic armamentarium. Because of their high toxicity the arsenicals are by no means the ideal chemotherapeutic agents. In the absence of better antitrypanosomal agents arsenicals are still being used. Melarsen oxide and Melarsoprol are representative members of the trivalent arsenicals and are currently being used for the treatment of bloodstream forms of West and East African trypanosomiasis. Another category of drugs effective against Chagas' disease is represented by Nifurtimox, a derivative of nitrofuran. The arsenicals are known for their inhibition of the trypanothione system. Nifurtimox and nitrofuran owe their antiparasitic effect to their ability to generate nitro-radicals that promote the formation of toxic oxygen metabolites inside the protozoal parasites. Both organic arsenicals and nitrofuran derivatives produce oxidative stress that cannot be compensated for by the antioxidant defenses of the trypanothione system.

MELARSOPROL

Inhibition of trypanothione reductase by the trivalent arsenicals occurs at the disulfide binding site with prior reduction of the redox-active disulfide bridge by NADPH as an essential prerequisite. Fairlamb and his associates reported that the active principle for the antitrypanosomal action of arsenicals is a

MELARSEN OXIDE

MELARSEN OXIDE - Try(SH) $_2$ adduct (**MelT**)

chemically stable adduct (inclusion complex) (Fairlamb, Henderson, & Cerami, 1989). When the trivalent arsenical melarsen oxide is used the adduct is a complex of reduced trypanothione and the arsenical, which is abbreviated MelT. Kinetic studies showed that the degree of inhibition by the adduct was decreased when the concentration of substrate $T(S)_2$ was increased, indicating that the inhibition is competitive. These results with MelT identified two possibilities for explaining the selective lethal effect of melarsen. First, the trivalent arsenical drug sequesters $T(SH)_2$ as a dithioarsane, depriving the parasite of its main sulfhydryl antioxidant. Second, inhibition of trypanothione reductase by the arsenical adduct deprives the parasite of the essential enzyme system that is responsible for keeping trypanothione reduced. It appears that both effects occur synergistically once melarsen oxide enters the trypanosome cell and the intracellular reducing environment of the parasite is lowered. Further kinetic studies showed that the affinity of reduced trypanothione for melarsen oxide was greater than that of other monothiols such as GSH or cysteine on which the mammalian host depends for maintaining a high thiol level. Thus the higher affinity of the arsenical to reduced trypanothione than to other monothiol compounds plays a role in the selectivity of arsenical action against trypanosomes as opposed to mammalian cells.

$$
\begin{array}{cc}
\text{CH}_2\text{OH} & \text{HOH}_2\text{C} \\
\text{CHOH} & \text{HOHC} \\
\text{CHO} \quad \text{OH} \quad \text{O}^- \quad \text{OHC} & 3\text{Na}^+ \\
\text{CHO-Sb-O-Sb-OHC} \\
\text{CHO} \qquad \text{OHC} \\
\text{COO}^- \qquad {}^-\text{OOC}
\end{array}
$$

SODIUM STIBOGLUCONATE
(PENTOSTAM)

Inhibition of Trypanothione Reductase from *Leishmania* by Trivalent Antimonials

For almost a century organic antimonials have been the drugs of choice against *Leishmania* infections of the intestine, skin, and mucous membranes. It is generally understood that the pentavalent antimonials such as Sodium stibogluconate (Pentostam) are only active following bioreduction to the trivalent Sb III form (Goodwin & Page, 1943). This is the main reason for using trivalent antimonials such as trivalent antimonyl sodium gluconate (Triostam) for *in vitro* research work. Like trypanosomes, the *Leishmanias* use trypanothione rather than glutathione. The intracellular level of dihydrotrypanothione is maintained by the NADPH-dependent trypanothione reductase. Trypanothione reductase from *L. donovani* has been purified to homogeneity and showed high specificity for trypanothione disulfide as a substrate (Cunningham & Fairlamb, 1995). Because of the chemical similarity between organic antimonials and arsenicals, the possibility was investigated by Cunningham and Fairlamb whether the tripanothione reductase is inhibited by organic antimonials. The trivalent antimonials, Triostam, potassium antimony tartrate, and antimony chloride, were all found to be inhibitors of trypanothione reductase. However, both trypanothione reductase from *L. donovani* and glutathione reductase from human erythrocytes are inhibited by Triostam. There is no significant quantitative difference between the inhibition of the two enzymes. The molecular mechanism for enzyme inhibition by the trivalent organic antimonials appears to be similar to that shown for trivalent arsenicals (Fairlamb *et al.*, 1989). Inhibition occurs within the disulfide-binding site after reduction of the redox-active disulfide bridge by NADPH. In fact, the adduct of melarsen oxide and dihydrotrypanothione, MelT, confers protection against Triostam inhibition, suggesting a similarity in the mechanism of inhibition. A transient mono-thio-stibane adduct is succeeded by a stable dithio-stibane complex that includes both redox-active cysteine residues at the active site (Cunningham & Fairlamb, 1995).

Tryparedoxins

Tryparedoxins (TXNs) are a newly identified group of enzymes present in parasitic *Trypanosomatidae* that are related to the host thioredoxins. TXNs catalyze the reduction of peroxiredoxin-type peroxidases by trypanothione. They are involved in hydroperoxide detoxification and also have some connection to the virulence of trypanosomes. Recently, Flohé and his associates identified structural and biochemical differences between these enzymes in *Crithidia fasciculata* and those of the host's thioredoxins (Hofmann *et al.*, 2001). The differences indicate that these enzymes may become important targets for discovery of antitrypanosomal agents (see also Epilogue).

Antiprotozoal Effects of Nifurtimox

Although dihydrotrypanothione/trypanothione reductase is the principal system for maintaining high thiol levels, trypanosomes lack catalase and have very low levels of peroxidase. The parasites are, therefore, vulnerable to by-products of O_2 reduction such as H_2O_2 and the superoxide anion O_2^-. Mammals are well endowed with catalases, peroxidases, and superoxide dismutases with the main function of catalyzing the dismutation of these highly reactive and destructive radicals.

NIFURTIMOX

Protozoal parasites as a group are deficient in defense mechanisms against oxygen toxicity. Consequently, a rational approach to their chemotherapy might take advantage of this important aspect of their biochemistry. Nifurtimox is a nitrofuran that is effective in the treatment of Chagas' disease in humans. This effect appears to be due to the generation of oxygen reduction products. The formation of a nitro anion radical as a result of reduction of nifurtimox followed by autoxidation of this radical results in the generation of the superoxide anion O_2^- and other oxygen by-products such as H_2O_2 and hydroperoxyl radicals. The effect of these highly reactive molecules appears to be responsible for the cytotoxic effect of nifurtimox as an antitrypanosomal agent (Docampo, 1990).

Polyamines and Ornithine Decarboxylase

Putrescine, spermine, and spermidine are polyamines that have a ubiquitous distribution. Although not all of their functions have been firmly identified, current experimental evidence indicates that there is a good correlation between levels of polyamines in cells and rapid growth during cell proliferation and differentiation. For a comprehensive review of the biochemistry and postulated functions of polyamines see Tabor & Tabor (1984). The polycations, spermidine and spermine, are thought to be involved in packaging and interacting with nucleic acids. Based on correlative evidence they appear to regulate DNA replication, transcription, and translation during cell proliferation. Since trypanosomes and malaria, like many other protozoal parasites, depend for their life on cyclical multiplication and differentiation, the possibility that polyamines could be attractive targets for antiprotozoal agents has been the subject of investigation.

Parasites depend on their own metabolic systems for the synthesis of polyamines. Ornithine serves as the principal precursor in their synthesis. The first reaction in this metabolic pathway is the decarboxylation of ornithine catalyzed by ornithine decarboxylase (ODC). Because ODC is the rate-limiting reaction in the pathway it has been purified and cloned from mammals, several other eukaryotes (*C. fasciculata, Trypanosoma,* and *Leishmania*) (Fairlamb & Cerami, 1992), and prokaryotes. The synthesis of the higher polyamines, spermidine and spermine, are successively carried out from putrescine by spermidine and spermine synthases. In these reactions there is successive addition of aminopropyl groups, which are derived from decarboxylated S-adenosyl methionine (dSAM). Experiments to prove that ODC is essential for survival of *T. brucei* in *in vitro* cultures were done by C. C. Wang and his associates. Deletion of the gene for ODC in knockout experiments using cloned *T. brucei* gave cells with no detectable enzyme activity. Growth of these cells in culture medium was dependent on the addition of putrescine, the product of the ODC reaction (Li *et al.*, 1996).

The activity of ornithine decarboxylase is controlled by two types of proteins. Enzyme activity is inhibited by one or more proteins called ornithine decarboxylase antizymes, which have a high affinity to bind to ODC noncovalently. The other control protein is an "activator" called "anti-antizyme." It has an affinity to the antizyme protein and it activates ODC by binding to the antizyme protein and thus nullifying its inhibitory effect by releasing the ODC from its complex with antizyme.

Ornithine Decarboxylase Inhibitors as Antiparasitic Agents

Because of the importance of ODC in the synthesis of polyamines, attempts were made to develop drugs that inhibit this enzyme. DL-α-difluoromethylornithine (eflornithine or DFMO) and its analogs were found to act as irreversible "suicide" (enzyme-activated) inhibitors of ODC (Bacchi *et al.*, 1980). They were originally thought to act as chemotherapeutic agents on cultured tumors as well as in laboratory animals. Initial experiments did not lead to successful treatment of human cancer (Sjoerdsma & Schechter, 1984). DFMO eventually became the lead compound as an inhibitor of ODC and was found to have antiprotozoal effects. From the clinical point of view DFMO is now highly recommended as one of the first-line treatments against *T. brucei* infections in humans. Usually melarsoprol is looked at as the first choice and DFMO as the back-up drug. Of 87 patients at the late stage of *T. brucei* infection that were treated with DFMO, only 8 relapsed one year after treatment (Milord *et al.*, 1992).

$$
\begin{array}{c}
H \\
| \\
F-C-F \\
| \\
HOOC-C-(CH_2)_3-NH_2 \\
| \\
NH_2
\end{array}
$$

EFLORNITHINE

Experimental results from treating *T. brucei* infected rats with DFMO showed that the infected animals were kept alive by DFMO whereas all control untreated rats died 72–80 hours after infection. These experiments were used to identify the mechanism of action of DFMO. Ornithine decarboxylase in animals that were treated with DFMO showed a 99% decrease in enzyme activity 12 hours after treatment. This marked decrease in enzyme activity was accompanied by a great decrease in putrescine and spermidine levels. The fact that ODC activity as well as putrescine levels were reduced indicate that ODC is the target for the effect of DFMO on the parasite (Bacchi *et al.*, 1983). Further evidence for implicating ODC and other polyamines in the mechanism of action of DFMO was that the antitrypanosomal action was antagonized when polyamines were added to the parasite cultures or co-administered with

DFMO to infected mice (Nathan *et al.*, 1981; Bacchi *et al.*, 1983; Sjoerdsma & Schechter, 1984).

DMFO was found to act as an inhibitor of ornithine decarboxylase because of its reaction at the active site of the enzyme. The inhibition is based on the mechanism of the catalytic reaction of the enzyme. The details of the mechanism of irreversible inactivation of ODC by DMFO were investigated using purified recombinant mouse ODC. Lysine 69, part of a heptapeptide, was identified at the ODC catalytic site where pyridoxal phosphate binds for ornithine decarboxylation. Instead of pyridoxal phosphate making a Schiff base with ornithine, a highly reactive PLP-DMFO adduct is covalently bound to the catalytic site of the enzyme. Another covalent adduct with DFMO was identified at cysteine 360 of the catalytic site of ODC. Both lysine and cysteine in the catalytic site of the enzyme were present in the active sites of all eukaryotic ODCs (Poulin *et al.*, 1992). Thus, DFMO appears to make use of the ODC catalytic mechanism to bind covalently to the enzyme and to kill the parasite. For a more complete description of the mechanism of action of DFMO see Nelson & Cox (2000).

Rate of Enzyme Turnover and Antiparasitic Action. One puzzle that was difficult to understand about the mechanism of antiparasitic action of DFMO was that the mammalian ODC is more sensitive to inhibition by DFMO ($K_i = 39 \mu$M) than is the parasite enzyme ($K_i = 130 \mu$M) (Sjoerdsma & Schechter, 1984; Bitonti *et al.*, 1985). Nonetheless, the inhibitor has a selective effect against the parasite but not the host. Cloning of the gene of ODC from *T. brucei* in the laboratory of C. C. Wang (Phillips, Coffino, & Wang, 1987) contributed to a better understanding of the mechanism of antitrypanosomal action of DFMO. The gene encoding ODC is present in a single copy in the parasite and has 61.5% homology with the mouse enzyme. One significant difference between the trypanosome and mammalian enzyme is the C-terminus. ODC from the mouse has a region (amino acids 423–449) that is rich in proline (P), glutamic acid (E), serine (S), and threonine (T) named the PEST region. It has been reported before that eukaryotic proteins that have PEST regions are subject to rapid intracellular degradation (Rogers, Wells, & Rechsteiner, 1986). The gene of trypanosome ODC does not have that region and therefore has a longer half-life. The short half-life of the mammalian enzyme compared to the long half-life of the trypanosome enzyme appears to be due to an intrinsic structural difference between the two enzymes. This is confirmed by experiments that show that the intracellular *T. brucei* ODC is very stable and undergoes no detectable turnover compared to the mammalian decarboxylase that has a half-life of only 10–30 min. The molecular basis for the lack of an inhibitory effect by DFMO on the decarboxylase in the host cells is that mammalian ODC has a rapid turnover. The inhibited enzyme in the host is constantly being replaced by *de novo* synthesized protein. In the case of the parasite, the decarboxylase has a much longer half-life, and the inhibited enzyme probably undergoes no detectable change during the relatively short life of the trypanosome. It is interesting also to note that differences in the sensitivity of different strains of trypanosomes to DFMO correspond well to differences in the turnover rate of ODC. For example, DFMO-tolerant *T. brucei rhodesiense* has a shorter half-life (4.3 hours) compared to DFMO-susceptible *T. b. gambiense* (18 hours) (Iten *et al.*, 1997). The effect of DFMO on trypanosomal ODC is the only known example where the antiparasitic effect of a chemical agent depends on the turnover rate of the target enzyme in the parasite.

Trypanothione is synthesized from spermidine, glutathione, ATP, and the synthetic enzymes glutathionyl spermidine synthase and trypanothione synthase. As a consequence of DFMO inhibition of ODC there is a marked decrease in spermidine and trypanothione levels and, therefore, a decrease in the level of dihydrotrypanothione. The reduction of $T(SH)_2$ contributes to the antiprotozoal effect of DFMO through inhibition of the antioxidant defense of the parasites. The double mechanism of DFMO action is a good example of several antiparasitic agents that affect more than one reaction. There is some pharmacological benefit to be gained from an agent that has multiple effects. In some cases it is difficult to verify the primary target of drug action.

Glycolysis: A Target for Antitrypanosomal Agents

The bloodstream forms of *T. brucei* depend on glycolysis for energy production. The reported differences between the properties of the glycolytic enzymes as well as their unique structural compartmentation in the glycosomes of the parasites make them an attractive target for designing antitrypanosomal agents. Trypanosomes deprived of glucose in *in vitro* culture would perish within a few minutes (Seyfang & Duszenko, 1991). Furthermore, phloretin, an inhibitor of membranous glucose transport, when added to cultured trypanosomes, is lethal to the parasites.

The Glycosome

In *T. brucei* the initial seven glycolytic enzymes that catalyze glucose to 3-phosphoglycerate are compartmented in a cellular organelle named a glycosome (Fig. 5.2). 3-Phosphoglycerate is further metabolized in the cytosol to pyruvate through the usual glycolytic enzymes. These trypanosomes in the bloodstream

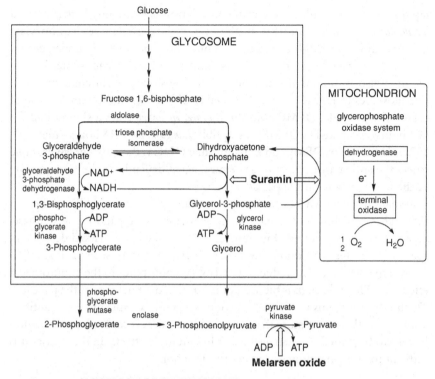

Figure 5.2. The glycolytic pathway of bloodstream forms of *T. brucei*. Shown are glycolytic enzymes and intermediates in the glycosome from fructose 6-bisphosphate that form 3-phosphoglycerate and glycerol. Net ATP synthesis is shown to occur in the cytoplasm as a product of pyruvate kinase. Under aerobic conditions NADH produced by glyceraldehyde 3-phosphate dehydrogenase is reoxidized by molecular oxygen in the mitochondria via the glycerophosphate oxidase system with the ultimate reduction of oxygen to water, oxidation of NADH to NAD, and the reconversion of glycerol 3-phosphate to dihydroxyacetone phosphate. This is catabolized via the triosephosphate isomerase to glyceraldehyde 3-phosphate and eventually to phosphoenolpyruvate and to pyruvate where an extra ATP molecule is produced. Sites of action of suramin and melarsen oxide are shown. Adapted from Bakker *et al.* (2000).

lack lactic dehydrogenase and the tricarboxylic acid cycle enzymes. Therefore, the glycolytic pathway ends with pyruvate as the main metabolic end product. Pyruvate produced by glycolysis is excreted into the host bloodstream by a facilitated diffusion process (Wiemer *et al.*, 1992). Along the cytosolic pathway of 3-phosphoglycerate to pyruvate ATP is synthesized by pyruvate kinase. Because the trypanosomes have no lactic dehydrogenase (necessary for the reoxidation of NADH to NAD) they have a special reaction to carry out this process. Under aerobic conditions NADH is reoxidized by molecular oxygen in the mitochondria. This is catalyzed by mitochondrial glycerophosphate oxidase. The same enzyme reconverts glycerol 3-phosphate to dihydroxyacetone phosphate, which is catabolized to pyruvate via glycosomal triose phosphate isomerase. Transfer of both glycerol 3-phosphate and dihydroxyacetone phosphate across the glycosomal membranes to reach the mitochondria appears to be carried out through a transporter in the glycosomal membrane. The biochemical nature of the transporter has still to be investigated. Under anaerobic conditions the mitochondrial glycerophosphate oxidase cannot function. Instead glycerol 3-phosphate is catalyzed to glycerol with the concomitant synthesis of ATP. This reaction is catalyzed by the reversed enzyme of glycosomal glycerol kinase. The metabolic balance of ATP production under aerobic conditions is two ATP molecules per one molecule of glucose. Under anaerobic conditions the net production is one mole of ATP per one mole of glucose utilized. This should be compared with the more efficient mitochondrial oxidative phosphorylation in the host cells of thirty-six molecules of ATP per one molecule of glucose. The poor yield of energy by the trypanosome is compensated for by a high rate of glycolysis, which has been reported to be 50 times higher than that in mammalian cells (Brohn & Clarkson, 1980). The compartmentalization of the glycolytic enzymes and the intermediate metabolites in the glycosome helps to maintain this high rate of glycolysis. In the glycosomes the enzymes are concentrated (340 mg protein/ml) and the steady-state concentration of the metabolites are at millimolar levels (Visser, Opperdoes, & Borst, 1981). Added to that is easy channeling between the metabolites and the enzymes in the glycosomes that facilitates the catalytic reactions. Also, constant availability of the blood surrounding the parasites makes it possible for constant absorption of glucose by facilitated diffusion into the parasites (Eisenthal, Game, & Holman, 1989). To keep up with their energy needs, the parasites have to consume large amounts of glucose, equal to their own dry weight every hour (Fairlamb, 1982).

Genes encoding several of the glycosomal enzymes from *T. brucei* have been cloned. In addition, studies on the control of glycolytic flux using computer simulation and kinetic properties of the enzymes and glucose transporters have shown that the control of glycolytic flux is shared by several enzymes and transporters in the glycolytic pathway (Bakker *et al.*, 2000). Results of these analyses provide quantitative information about the effects of inhibition of individual enzymes on the glycolytic flux. The most promising target in the glucose pathway of trypanosomes appears to be the plasma membrane glucose transporter. This is followed by a group of four enzymes: aldolase, glyceraldehyde 3-phosphate dehydrogenase, phosphoglycerate kinase, and glycerol 3-phosphate dehydrogenase. The kinases, hexokinase, phosphofructokinase, and pyruvic kinase, which are considered to be rate-limiting steps in many systems, were not rate limiting in the trypanosome glycosomes.

Glucose Transporters

Because of the important role that the glucose transporters play in the control of glycolysis in trypanosomes, their structures and functions are being investigated. The transporters are protected from the host immune system by their location under the parasite surface coat that consists of variant surface glycoproteins (VSGs). For a comprehensive review see Borst & Fairlamb (1998). A high rate of glucose absorption is facilitated in the bloodstream form of *T. brucei* by glucose transporter THT_1. Similar glucose transporters have also been identified in *T. cruzi* and *T. vivax*. Many of these transporters have been cloned and expressed in bacteria or *Xenopus*. The trypanosome transporters have twelve transmembrane regions similar to those of the mammalian erythrocyte transporters $GLUT_1$. Drugs that are currently in use against trypanosomes have been tested on these expressed transporters, but none showed a selective inhibition of the transporters (Barrett *et al.*, 1998). This group is preparing sugar analogs with chemical substituents at the hydroxyl positions 2 and 6 of the glucose ring which would not interfere with binding to the glucose transporter in *T. brucei*.

Pyruvate Transporters

In spite of the evidence that glycolytic enzymes and glycosome membrane transporters should be considered as prime targets for discovering less toxic and more effective antitrypanosomal agents there has been only limited success, but encouraging findings in the field are emerging. For example, studies on the pyruvate transporter across the membrane of the bloodstream form of *T. brucei* indicate that the carrier is a possible target for chemotherapeutic intervention (Wiemer, Michels, & Opperdoes, 1995). The transporter plays an important role in preventing acidification of the cytosol by facilitating the transport of pyruvate anion out of the parasite. Selectivity of the trypanosome transporter to inhibitors was demonstrated. The pyruvate transporter was found to be rather insensitive to inhibition by an α-cyano-4-hydroxycinnamate, a pyruvate analog that inhibits pyruvate transport across the plasma membrane in different eukaryotic cell types. In contrast, pyruvate transporters of *T. brucei* could be completely blocked by UK 5099 [α-cyano-β-(1-phenylindol-3-yl)acrylate] *in vitro*. Inhibition of pyruvate efflux from the cells led to cytosolic acidification of the parasites, inhibition of respiration, and swelling of the cells because of pyruvate retention. Eventually the parasites form clumps of dead cells. These preliminary experiments confirm that pyruvate export is carrier mediated and that it is essential for survival of the trypanosomes. In addition, the fact that these transporters are located in the plasma membrane of the trypanosomes, a site that is readily accessible to drugs given through the circulation, makes them an important prospective target for antitrypanosomal drugs.

α-Glycerophosphate Oxidase

Salicylhydroxamic acid (SHAM), an iron chelator, was identified as an inhibitor of α-glycerophosphate oxidase. Such inhibition stops the mitochondrial contribution to the glycolytic pathway. The consequence is that the parasite would be dependent on producing only one mole of ATP per mole of glucose as under anaerobic conditions. Inhibition by SHAM can be enhanced by simultaneous administration of glycerol, which inhibits reversal of the glycerol kinase-catalyzed reaction. Increasing the level of glycerol reduces the reverse reaction by mass action. When glycerol is present in excess, ADP phosphorylation to ATP through the glycerol kinase reaction is inhibited. Using combined SHAM–glycerol chemotherapy suppressed parasitemia rapidly in animal models. However, there was animal relapse after the administration of glycerol and SHAM. This has been attributed to inadequate tissue concentrations of the drugs or to distribution of trypanosomes to different parts of the host. For a review see Fairlamb (1982).

Other Glycolytic Targets

Whereas the bloodstream forms of *T. brucei* depend on glycosomal enzymes for their energy, *T. cruzi* live in human host cells and their metabolism is mainly aerobic using their mitochondria (Fairlamb, 1982). These organisms, as well as promastigotes of different species of *Leishmania*, depend on aerobic mitochondrial acetate:succinate CoA transferase for production of acetate from acetyl CoA with the ultimate formation of ATP from ADP (Van Hellemond, Opperdoes, & Tielens, 1998). Although the use of glycolytic enzymes as targets in these organisms may not appear to be rational because of the availability of energy-rich aerobic metabolism, Gelb and his associates have found that some of the glycolytic enzymes are sufficiently different from homologous mammalian enzymes that they can be good targets to inhibit energy production in these parasites. Glyceraldehyde 3-phosphate dehydrogenase (GAPDH) from *T. brucei*, *T. cruzi*, and *Leishmania* were reported to be structurally similar, but they have significant differences from the homologous human enzyme. Gelb and his associates have identified structure-based adenosine analogs that can selectively inhibit these trypanosomatid dehydrogenases but not the human enzyme (Suresh et al., 2001). By analyzing the crystal structure of L. mexicana GAPDH in complex with different adenosine analogs they found that those analogs containing substitutions on N-6 of the adenine ring and on the 2' position of the ribose moiety had the highest affinity to the enzyme and could also block parasite growth in vitro. Some of these adenosine analogs were tested for their inhibitory effects against phosphoglycerate kinase, GAPDH, and glycerol 3-P dehydrogenase (GPDH) from T. brucei. All three enzymes were inhibited. The best inhibitor in this series is 2-amino-N(6)-[2″(p-hydroxyphenyl)ethyl]adenosine, which inhibits growth at a low micromolar range (Bressi et al., 2000). A similar inhibitory effect against GPDH from the intracellular form of T. cruzi was also demonstrated with these nucleoside analogs. These results underscore the importance of using structure-based drug design to transform a lead compound with poor potency (IC_{50} for adenosine $\sim$ 50 mM) into several more potent inhibitors (Aronov et al., 1999).

SURAMIN SODIUM

Suramin and Glycolysis

Suramin is a polyanionic antiprotozoal agent that is particularly effective against T. brucei. The drug is chemically related to trypan blue and trypan red, two dyes that were originally investigated as antitrypanosomal agents by Paul Ehrlich. Their mechanism of action has been investigated by the groups of both Fairlamb and Opperdoes (Fairlamb & Bowman, 1980; Willson et al., 1993). Suramin appears to owe its antitrypanosomal effect to having a net negative charge at physiological pH. Suramin has two symmetrical, polysulfonated naphthylamine groups, one at each end of the molecule, which are 4 nm apart. Suramin interacts with several glycosomal enzymes that are positively charged at distances that are equivalent to those of the anionic groups in suramin. Fairlamb and Bowman (1980) reported that after suramin was injected into rats infected with T. brucei it was taken up slowly by the trypanosomes from the host. Once within the parasites suramin progressively inhibited both respiration and glycolysis. Suramin appears to inhibit all the trypanosome glycolytic enzymes more effectively than it does those of the host. However, the *in vivo* action of suramin on trypanosomes is due to a combined action against glycerol 3-phosphate dehydrogenase (a glycosomal enzyme) and glycerol 3-phosphate oxidase (in the mitochondria) (Fig. 5.2).

Melarsoprol

Melarsoprol and melarsen oxide trivalent organic arsenicals that affect the trypanothione system in trypanosomes, also inhibit several enzymes of carbohydrate metabolism in trypanosome homogenates. This is expected since many of these enzymes have sulfhydryl groups that have a high affinity to react with arsenicals. When given to intact parasites in culture, pyruvate kinase is selectively inhibited, as evidenced by the accumulation of phosphoenol pyruvate, the substrate of the enzyme (Fairlamb, 1982). Other glycolytic enzymes that were reported to be inhibited include phosphofructokinase and fructose-2,6-bisphosphatase (Wang, 1995). Verification that these enzymes are primarily involved in the antiparasitic effect of melarsoprol is lacking.

Mitochondrial DNA as a Target for Antitrypanosomal Agents

Mitochondrial DNA of trypanosomes and related protozoan parasites is usually referred to as kinetoplast DNA (kDNA). kDNA is a network of DNA that is made up of 5000 minicircles and a few dozen maxicircles. One characteristic feature of kDNA that makes it unique in nature is that all these circles are linked together into a single giant network. Thus, a single mitochondrion contains only one network of DNA. The functional rationale behind having this unusual arrangement of the DNA remains a subject of investigation. It is, however, recognized that maxicircles encode ribosomal RNAs and proteins involved in mitochondrial energy transduction such as cytochrome oxidase or NADH dehydrogenase. The function of the minicircles is to encode RNAs, which control editing specificity (Englund *et al.*, 1995). Replication of kDNA involves many topological interconversions. Each minicircle is first decatenated (unchained from the network) before undergoing several topological interconversions that are catalyzed by topisomerases. These are enzymes that catalyze the interconversion of different topological isomers of DNA.

Several antitrypanosomal agents have been reported to bind kDNA and disrupt its function (Shapiro & Englund, 1990; Shapiro, 1994). These include pentamidine, ethidium bromide, and samorin (isometamidium chloride), the last of which has the structural features of ethidium bromide. Although details of the molecular action of these agents remains to be elucidated there is a general agreement that the earliest site of action at therapeutic drug levels is characterized by disruption of kDNA in the trypanosomes. Trypanosomes treated with these agents develop dyskinetoplastic cell populations. These are cells with mitochondrial membranes but no detectable kDNA. Recent investigations by Shapiro and her associates showed that all the above mentioned agents bind to and disrupt kDNA structure. Experimental evidence also indicates that the minicircles linearized by the antitrypanosomal agents have bound proteins that, together with kDNA, give characteristic features of type II topoisomerase-DNA complexes. This group of topoisomerases serve as good candidates for drug targets. They are molecular targets for several antibacterial and antitumor agents because of their effect on DNA replication and transcription. Topoisomerase I catalyzes single-stranded breaks in DNA whereas type II catalyzes double-stranded breaks.

Purine and Pyrimidine Metabolism as Targets

Purine and pyrimidine nucleotides play an essential role in the survival of all living organisms. They are the basic biochemical units of both DNA and RNA and also participate in many other biosynthetic processes. Protozoal parasites in particular depend for their survival on rapid multiplication in the host environment. Consequently, there is a continual need for DNA and RNA synthesis by the parasites.

The relevance of studies on purine and pyrimidine metabolism in the search for selective antiparasitic agents became more important when fundamental differences between parasites and their mammalian hosts were revealed. During the past two decades there has been accumulating evidence to indicate that although mammalian hosts are able to synthesize their purines *de novo* from simple precursors, all parasites that have been studied have to utilize purines from their host environment only through the salvage pathways. This is probably an evolutionary metabolic step for the parasite to make use of the host's nucleic acid degradation products with great savings to its own energy resources. This fundamental difference between parasite and host is now the basis of a rational approach to select for chemotherapeutic agents that are toxic to the parasite and not the host. However, all parasites with the exception of the amitochondrial protists *Giardia lamblia, Trichomonas vaginalis,* and *Tritrichomonas foetus* are able to synthesize pyrimidines *de novo* as well as use the salvage pathway. Because of the similarity between the pyrimidine metabolism in most of the parasites and that of the host, targets among reactions from these pathways may not have the same specificity as those in the salvage pathways. Therefore, the *de novo* pathways for pyrimidine metabolism have not been extensively considered in a strategy for finding new antiparasitic agents.

Purine Salvage Pathways

In mammals and many parasitic organisms purine bases are produced by the hydrolysis of nucleic acids or nucleotides by different phosphatases and nucleotidases. The purines are metabolized to nucleotides through the salvage reactions. An important intermediate is the activated sugar unit 5-phosphoribosyl-1-pyrophosphate (PRPP). Purine bases can be salvaged by the transfer of the ribose phosphate moiety of PRPP to form the corresponding ribonucleotide. Two salvage enzymes that have different substrate specificities are the adenine phosphoribosyl transferase (APRT), which catalyzes the synthesis of AMP, and hypoxanthine-guanine phosphoribosyl transferase (HGPRT), which catalyzes the synthesis of inosinate (IMP) or guanylate (GMP):

$$\text{adenine} + \text{PRPP} \xrightarrow{\text{APRT}} \text{adenylate} + \text{PP}_i$$

$$\text{hypoxanthine} + \text{PRPP} \xrightarrow{\text{HGPRT}} \text{inosinate} + \text{PP}_i$$

$$\text{guanine} + \text{PRPP} \xrightarrow{\text{HGPRT}} \text{guanylate} + \text{PP}_i.$$

Individual phosphoribosyl transferases (PRTs) catalyze the salvage of different purines by reacting with PRPP to yield nucleoside-5'-monophosphates. Several reactions are involved in side metabolic interconversions in both humans and parasites. A comprehensive review of purine salvage pathways in mammals and in different parasites was written by Berens, Krug, & Marr (1995). A schematic diagram of some purine pathways in mammals and kinetoplastids is shown in Fig. 5.3. Differences in parasite salvage pathways appear to be dictated by the purine composition of the parasites' habitats in the host. Each genus of parasite appears to have its own type of PRT with characteristic substrate affinities and kinetic properties. For example, *L. donovani* has three distinct PRTs, one for adenine, a second for hypoxanthine and guanine, and a third for xanthine (Tuttle & Krenitsky, 1980). Xanthine PRT appears to be unique to the *Leishmanias* whereas mammals have enzymes comparable to APRT and HGPRT. Xanthine PRT has been cloned and biochemically characterized by Jardim *et al.* (1999). HGPRT has been identified in *T. brucei* and *T. cruzi*. *Tritrichomonas foetus* (a protozoa that infects cattle and causes infertility or abortion) has a

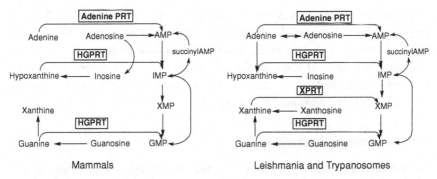

Figure 5.3. Schematic diagram of purine salvage and some interconversion pathways in mammals and kinetoplastids. Adenine PRT, adenine phosphoribosyl transferase; HGPRT, hypoxanthine-guanine phosphoribosyl transferase; XPRT, xanthine phosphoribosyl transferase. Adapted from Berens *et al.* (1995).

hypoxanthine/guanine/xanthine PRT (HGXPRT). This enzyme has also been identified in several *Plasmodia* species (Queen, Vander Jagt, & Reyes, 1988). Collateral reactions could involve conversion of a nucleoside into a nucleotide such as the phosphorylation of adenosine to AMP by a nucleoside kinase. Most cells interconvert adenine and guanine nucleotides using IMP and AMP as intermediates (Fig. 5.3). There is good qualitative, but not quantitative, similarity between the metabolic sequence for purine metabolism among various morphological forms of *Trypanosoma* and *Leishmania*.

Because of their critical role in purine salvage pathways in parasites and the predictions that PRTs are potential antiparasitic targets, more molecular and biochemical studies are currently being done to examine these enzymes and to make them available in milligram amounts for crystallographic studies. Genes encoding HGPRT proteins were cloned from seven parasitic protozoa including *T. brucei, T. Cruzi,* and *L. donovani* (Ullman & Carter, 1995, 1997). Information regarding the crystal structure of PRT enzymes is needed for designing potential antiparasitic agents. The crystal structure of human HGPRT is available (Eads *et al.*, 1994) and should be of assistance in identifying the molecular basis of selectively active antiparasitic agents. Using the atomic coordinates of the human HGPRT obtained by X-ray crystallography Ullman's group was able to construct 3-D molecular models of HGPRT from *T. cruzi, T. Brucei,* and *L. donovani.* Differences between the structures of host and parasite enzymes could be utilized for the design of new, more effective, and less toxic agents. Another step toward that goal has recently been reported. HPRT from *T. cruzi* was crystallized with the inosine analog Formycin B and the crystal structure was determined to 1.4 Å, the highest resolution reported for a PRT (Focia *et al.*, 1998).

Nucleoside Transporters

Parasites have a mechanism by which they can salvage nucleotides from the host. Nucleotides have first to be dephosphorylated to nucleosides for easy passage across the protozoal surface membrane. Transporters that carry different nucleosides have been described in several parasites. For example, adenosine appears to be the preferred substrate for salvage by intracellular *T. gondii* tachyzoites. A plasma membrane transporter that preferentially transports adenosine was described by Schwab *et al.* (1995). Once adenosine enters the tachyzoites it is rapidly converted to AMP, ADP, and ATP by adenosine kinase. The same transporter can also transport inosine, hypoxanthine, and adenine. Adenosine, however, is the preferred purine substrate for salvage by this transporter and it therefore plays a crucial role in the intracellular survival of these parasites. Most protozoal parasites have two or three separate nucleoside transport pathways. *T. brucei* has two types of adenosine transporters: a P_1 type, which also transports inosine, and a P_2 type, which also transports adenine. The adenosine P_2 transporter was also reported to transport melarsen oxide, an arsenical-chemotherapeutic agent effective against African sleeping sickness (Carter & Fairlamb, 1993). Carter and Fairlamb showed that in melarsen-resistant strains the P_2 transporter is not functional and therefore the parasite is unable to take up the drug. In support of this finding dipyridamole, a general nucleoside transporter inhibitor, was found to inhibit the uptake of melarsen when added to cultures of sensitive trypanosomes and thus antagonized the lysis of the arsenical-treated parasites. The results support the suggestion that resistance to melarsen is due to a lack in the ability of the parasites to transport this arsenical compound. A comparison among the structures of adenosine, melarsen oxide, and dipyridamole indicates that the melamine moiety that is shared by these three compounds could provide the structural basis for directing these molecules to the parasites' P_2 transporter receptors. Many N^6-nitrobenzylthiopurine derivatives

ADENOSINE   MELARSEN OXIDE   DIPYRIDAMOLE
(Nucleoside transporter inhibitor)

have been reported to be potent inhibitors of nucleoside transporters in various types of mammalian cells. Good examples are nitrobenzylthioinosine, nitrobenzylthioguanosine, and dipyridamole. These compounds have been reported to have high affinity to specific nucleoside transporters in mammalian cells with K_ds in nanomoles (Plagemann, Wohlhueter, & Woffendin, 1988). A rational strategy for finding chemotherapeutic agents could include the use of inhibitors that can selectively bind to the parasite transporters. Investigations in this field have not exploited these targets as yet.

Nucleoside Ribohydrolases

Nucleoside ribohydrolases are a group of enzymes that are involved in purine salvage pathways in African trypanosomes and other protozoa, yeast, and some bacteria but are not found in mammals. These enzymes catalyze the hydrolysis of the N-riboside bond between N-9 of the purine and C-1′ of the (deoxy-)ribose:

$$(deoxy\text{-})nucleoside + H_2O \longrightarrow base + (deoxy\text{-})ribose.$$

A ribohydrolase that is inosine-adenosine-guanosine preferring (IAG) has been identified and cloned from *T. brucei brucei* (Pelle *et al.*, 1998). Southern blot analysis identified the gene in other African trypanosomes, and to a lesser degree in *T. cruzi*. Recent studies on the reaction mechanisms and transition states of these enzymes have contributed to the design of specific inhibitors of the parasite nucleoside hydrolases (Miles *et al.*, 1999). Novel transition-state analog inhibitors of IAG hydrolase from African trypanosomes have been identified. The most potent is a 1-substituted iminoribitol, which inhibits the IAG hydrolase at subnanomolar concentrations. Because of the high potency of these inhibitors they deserve further examination for trypanosome survival both *in vitro* and *in vivo*.

Purine Analogs as Antiparasitic Agents

The importance of the purine salvage pathway as a basis for chemotherapy has been realized from studies on allopurinol (4-hydroxypyrazolo-[3,4-D] pyrimidine) and derivatives by (Marr, 1991). Allopurinol is an isomer of hypoxanthine, which was shown to have significant therapeutic effects in patients

HYPOXANTHINE   ALLOPURINOL

suffering from cutaneous leishmaniasis as well as in patients with *T. cruzi* infections. Allopurinol per se is not pharmacologically an inhibitor, but it acts as a "subversive substrate." The hypoxanthine-guanine PRT of *Trypanosoma* and *Leishmania* catalyzes the conversion of allopurinol to the IMP analog. Allopurinol also provides a typical example of how a false substrate can be metabolized by the enzymes of the parasite to false metabolic products that inhibit vital pathways in the parasite. Allopurinol nucleoside is used clinically for the treatment of visceral leishmaniasis. It is converted to the same nucleotide analog of IMP. The subversive substrates are converted to the AMP derivative by the parasite adenylosuccinate lyase and adenylosuccinate synthase. They are then converted to the ATP derivative by two phosphorylation steps and finally are incorporated into RNA, where they cause breakdown of mRNA. This results in inhibition of protein synthesis. The toxic effects on purine synthesis and on protein synthesis both account for the antiparasitic action of allopurinol. These effects are common to all hemoflagellates (e.g., *Leishmania* and *Trypanosomes*; Marr & Ullman, 1995). The biochemical effects of allopurinol on purine metabolic pathways have been demonstrated in enzymatic reactions and in tissue culture as well as in patients suffering from cutaneous leishmaniasis. Treatment with orally administered allopurinol gave a cure rate of 80% of infected patients (Martinez & Marr, 1992). Patients suffering from *Trypanosoma cruzi* had the same cure rates as those taking nitrofurans with little or no toxicity (Marr, 1991). Based on the reported clinical and biochemical effects against the parasite's purine salvage pathway it appears that allopurinol and its analogs are lead compounds for a rational search for more effective suicidal substrates as antiparasitic agents.

FORMYCIN B   FORMYCIN A monophosphate

Formycin B is an inosine analog that was reported to inhibit the *in vitro* growth of the promastigote forms of several *Leishmania* species. In addition, the drug is therapeutically active in hamsters infected with *L. donovani*. The mechanism of its antiparasitic action appears to be similar to that of allopurinol. The initial reaction is the formation of Formycin B monophosphate catalyzed by the *Leishmania* nucleoside phosphotransferase. This is the selective step in the parasites' metabolic machinery since mammalian cells do not phosphorylate Formycin B. The monophosphate is then converted to Formycin A monophosphate by the adenylosuccinate synthase/lyase. Formycin A monophosphate is further phosphorylated by AMP kinase and nucleoside diphosphokinase to Formycin A triphosphate. These cytotoxic adenosine nucleotides of Formycin (suicidal substrates) finally become incorporated into RNA. The cytotoxic effect of Formycin B is primarily due to the accumulation of toxic metabolites of Formycin A in the parasite (Berman, Rainey, & Santi, 1983; Rainey & Santi, 1983).

Several protozoal parasites including *Plasmodium falciparum*, *Eimeria tenella*, *Tritrichomonas foetus*, and *Toxoplasma gondii* were reported to have enzymes that can accept xanthine as well as hypoxanthine and adenine as substrates: hypoxanthine-guanine-xanthine phosphoribosyl transferase (HGXPRT) (Ullman & Carter, 1995). In contrast, the PRT in the human host can only utilize adenine and hypoxanthine (HGPRT). Because of the differences in the substrate specificity of host and parasite enzymes xanthine analogs (8-azoxanthine or 6-thioxanthine) could only be metabolized by the parasites. A newly discovered xanthine phosphoribosyltransferase (XPRT) in *L. donovani* that has no human homolog has been reported recently. The enzyme preferentially phosphoribosylates xanthine. Guanine and hypoxanthine could also be recognized by the enzyme but had high K_m values. The enzyme is a good target for evaluating novel xanthine analogs against *Leishmania* (Jardim *et al.*, 1999). These analogs act as subversive substrates when they become incorporated into the nucleotide pool of the parasites and produce cytotoxic effects. The selectivity of some of the HGPRTs and XPRTs for the xanthine analogs and the fact that some of the HGPRTs have been isolated and well studied make them attractive targets for identifying antiparasitic agents.

The suitability of inhibitors of *T. gondii* HGXPRT as preferred targets to block the salvage pathways has been dampened by a recent report that knockout mutants lacking the enzyme grow as normally as the wild type *in vitro*. Although this finding has to be further investigated it still raises the possibility that other enzymes in the parasite are involved in a bypass of the salvage pathway (Donald *et al.*, 1996).

In *T. gondii* adenosine has a different metabolic pathway compared to that of the mammalian host. In the host, adenosine is first deaminated by adenosine deaminase to inosine, which is then cleaved to hypoxanthine by nucleoside phosphorylase. These enzymes are not present in significant amounts in *T. gondii* (Krug, Marr, & Berens, 1989). However, *T. gondii* has ten times the adenosine kinase activity when compared to other purine salvage enzymes. Adenosine appears to be the source of nucleobases or nucleosides in *T. gondii*. High adenosine kinase activity in the parasite results in the direct conversion of adenosine to AMP, from which other purine nucleotides are synthesized. These differences in the salvage pathways for adenosine in the parasite compared to those of the host indicate that adenosine kinase could be a good target for selecting antiparasitic agents against *T. gondii*. An important contribution toward the use of adenosine kinase as a target in *T. gondii* has recently been reported. Sullivan *et al.* (1999) cloned the enzyme and have verified its function in knockout mutant parasites. Adenosine kinase cDNA under the control of a heterologous promoter expressed adenosine kinase activity in the mutant parasites (Sullivan *et al.*, 1999). The cDNA is now available for experimentation to help in identifying more selective inhibitors.

Recent studies on the structure–activity relationship for the binding of some 128 purine nucleoside analogs to adenosine kinase of *T. gondii* provide some support for this approach (Iltzsch *et al.*, 1995). For example, two purine nucleoside analogs, N^6-(*p*-methoxybenzoyl) adenosine and 7-iodo-7-deazaadenosine (iodotubercidin) showed greater affinity to the adenosine kinase from *T. gondii* than the natural substrate, adenosine. Structure–activity relationships for the binding of nucleoside ligands to adenosine kinase predicted a "pocket" in the catalytic site of *T. gondii* adenosine kinase adjacent to the 6-position of adenosine that can accommodate large substituents. Crystallographic studies should be used to verify this interesting possibility and extend its use to identifying more selective antiparasitic agents.

Pyrimidine Metabolism Pathways as Possible Targets

Although *T. gondii*, like most other protozoal parasites, can fulfill its pyrimidine requirements from either *de novo* synthesis or the salvage pathways, present information indicates that the parasite differs from its mammalian host as to the steps of the pyrimidine salvage pathway (Iltzsch *et al.*, 1995; el Kouni *et al.*, 1996). These differences offer opportunities to identify new targets for selective antiparasitic chemotherapy. One of the reactions in the pyrimidine salvage pathway of *T. gondii* that is quite different from that of the host is the phosphorylation of pyrimidine nucleosides. Unlike the mammalian host the parasites lack pyrimidine nucleoside kinases (Iltzsch, 1993) and therefore cannot phosphorylate pyrimidine and deoxyribosides to their respective nucleoside 5′-monophosphates. The pyrimidine nucleosides are converted to uracil or thymine by a single nonspecific uridine phosphorylase. Uracil is then converted by a specific uracil phosphotransferase to UMP, from which all pyrimidine nucleotides can be synthesized (Iltzsch, 1993). Research is being conducted to impair or circumvent the pyrimidine salvage pathway. Because the uridine phosphorylase catalyzes the key reaction in the biosynthesis of pyrimidine nucleotides el Kouni and his associates (el Kouni *et al.*, 1996) examined the effect of inhibitors or "subversive substrates" on the enzyme. They studied the effect of changes in the uridine structure, including modifications to the uracil and pentose moieties and the binding of nucleoside ligands to uridine phosphorylase from *T. gondii*. Some potent compounds were discovered, including 5-(benzyloxybenzyl)-2,2′ anhydrouridine, which has a K_i value at the nanomolar level. Many of the chemical differences that influenced the binding to the parasite enzyme compared to the host enzyme should be advantageous in the design of selective inhibitors or "subversive" substrates for *T. gondii* uridine phosphorylase.

Pteridin Reductase in *Leishmania major* as a Target

Leishmania is endowed with a pathway for salvaging pteridines and folic acids from the host. They lack the pathway for the biosynthesis of pteridine that was described for malaria in Chapter 4. A newly discovered enzyme, pteridine reductase (PTR1) appears to play an important role in the salvage of unconjugated pterins (Nare *et al.*, 1997). PTR1 is an NADPH-dependent enzyme that catalyzes the reduction of biopterin to dihydrobiopterin and subsequently to tetrahydopterin. In addition to reducing biopterin, PTR1 can also catalyze the two-step reduction of folate to tetrahydrofolate, which is primarily catalyzed by the dihydrofolate reductase component of the bifunctional enzyme dihydrofolate reductase-thymidylate synthase (DHFR-TS). This enzyme is the target for the selective DHFR inhibitor methotrexate. The finding that PTR1 can reduce the folates as well as the biopterins indicates that PTR1 can function as a bypass of DHFR-TS when the latter is inhibited by methotrexate. Overexpression of PTR1 in resistant strains of *Leishmania* can account for the resistance to methotrexate (Callahan & Beverely, 1992). That investigation drew attention to PTR1 as an enzyme system that compromises the effect of antiprotozoal agents that inhibit DHFR-TS. Thus, to shut down the folate system that is essential for survival of *Leishmania*, both PTR1 and DHFR-TS have to be inhibited. Selective inhibitors to PTR1 or a single inhibitor that acts on both enzymes would constitute a rational approach to select for new antileishmanial agents. Beverley's group has used this strategy (Hardy *et al.*, 1997; Nare *et al.*, 1997). From a collection of pteridines, pteridine analogs, and pyrimidines potent inhibitors of *L. major* PTR1 and DHFR-TS were selected. Of these, only one compound, 2,4-diamino-6,7-diisopropylpteridine, was found to inhibit both enzymes and was capable of inhibiting *L. major* amastigotes in cultured macrophages (Hardy *et al.*, 1997).

This pteridine analog, however, appears to have no activity in animal infection models (Nare *et al.*, 1997). This is not unusual for a first trial. With the availability of the crystal structure of PTR1 determined by the Beverley and Hunter groups further screening of better inhibitors can now be done (Gourley *et al.*, 1999).

PTR1 is also present in *T. cruzi* as reported by Gamarro and his colleagues (Robello *et al.*, 1997). The trypanosome enzyme interaction with other enzyme systems has yet to be fully investigated.

Antigenic Variation in African Trypanosomes and the Design of Antiparasitic Agents

During their bloodstream life cycle, African trypanosomes are endowed with a layer of glycoproteins that cover their surface and are known as variant surface glycoproteins (VSG). The parasites use these proteins to evade the immune system by periodically expressing different sets of VSG proteins in succession. As soon as the host can make antibodies to one set of VSGs the parasite produces another. *T. cruzi* do not have the VSG because their exposure to the host's antibodies is limited and the parasites appear to have no need for protection against the host's immune system. VSG proteins in African trypanosomes are tethered to the membrane of the protozoa by an unusual glycosyl phosphatidylinositol (GPI) anchor (Ferguson & Cross, 1984). The main biochemical structure of this anchor, ethanolamine-P-Man_3-GlcN-phosphatidylinositol, is bound to the plasma membrane through covalently linked myristic acid (a saturated C_{14} fatty acid). The lipid component is made of dimyristoylglycerol and is unique to the trypanosomes. One strategy that has been tried is to see whether analogs of myristic acid could be selectively toxic to trypanosomes (Doering *et al.*, 1991). Analogs of myristate with oxygen substituted for one methylene group were examined in cell-free extracts and in intact organisms in culture. An analog with oxygen substitution at the eleventh carbon 11-oxatetradecanoic acid, termed O-11 [i.e., 10-(propoxy)decanoic acid] was found to be incorporated into GPI/myristate formation. This was found to substitute more efficiently than the

MYRISTATE

11-OXATETRADECANOIC ACID (O-11)
also known as 10-(PROPOXY)DECANOIC ACID

natural substrate (myristate) into glycolipid A (the GPI myristoylation precursor). Myristate analog O-11 was also shown to be toxic to trypanosomes cultured *in vitro* with an LD_{50} of 1 μM whereas at 200 μM the compound was not toxic to cultured mammalian cells. O-11, when incubated with trypanosomes *in vitro*, was incorporated into glycolipid A and VSG. This shows that the toxic effect is selective for the trypanosome because of its interference with GPI/myristate formation that is essential for anchoring the VSGs. Furthermore, the Englund group examined 244 more fatty acid analogs for incorporation into VSGs and their effects against trypanosomes. These studies showed correlation between the incorporation of oxatetradecanoic acid derivatives into the surface glycoproteins and toxicity to the parasites (Doering *et al.*, 1994). Unfortunately, three of the most toxic analogs of myristate and myristoyllysophosphatidylcholine, when tested in mice, were ineffective in curing the animals. This has been attributed to the possibility of rapid metabolism by the host (Werbovetz, Bacchi, & Englund, 1996). Further attempts to prepare more stable analogs may be worthwhile.

It had long been assumed that trypanosomes could not synthesize their own myristate or any other fatty acids but depended on the host's blood for their needs. One ambiguity about this assumption is that these parasites need more myristate than is available for salvage from the host's blood. Recently, Englund's group resolved that problem by establishing that the trypanosomes can synthesize part of their needed myristate (Morita, Paul, & Englund, 2000; Paul *et al.*, 2001). Myristate and its derivatives can be obtained by the parasites through diffusion across the membrane or through fatty acid transporters. Since the myristate *de novo* synthesis system and the mechanism to salvage from the host's blood are both unique to the parasite, both processes have been considered as prospective targets. Similarly, the process of GPI myristoylation could also be a potential target.

Attempts are also being made to identify inhibitors of the fatty acid synthesis in the trypanosomes. The synthetic pathway appears to be unique, particularly since myristate, the major product of the system, is preferentially incorporated into GPI. Thiolactomycin is an antibiotic that is known to inhibit bacterial, but not eukaryotic, fatty acid synthesis. This antibiotic was reported to inhibit trypanosomal fatty acid synthesis as well as cell growth in cultures at micromolar concentrations (Morita *et al.*, 2000). The expectation is that other antibiotics with similar properties will be investigated.

Several fatty acid biosynthetic enzymes have been identified in parasites of the phylum Apicomplexa, *Plasmodium falciparum*, and *Toxoplasma gondii*, all of which showed similar DNA sequence to the homologous enzymes in *T. brucei* (Waller *et al.*, 1998). As is the case with *T. brucei*, thiolactomycin inhibits the growth of *P. falciparum*. It appears that fatty acid biosynthesis plays an important role in malaria and in *T. gondii* and it may very well be a good target for antiparasitic drugs.

Protein Farnesylation as a Drug Target

Prenylation of proteins is another way for proteins to anchor to membranes. The process involves covalent attachment of prenyl groups such as 15-carbon farnesyl or 20-carbon geranylgeranyl groups to a protein. Three enzymes have been identified in different eukaryotic cells that prenylate proteins: protein farnesyl transferase (PFT), and protein geranylgeranyl transferases-1 and -2 (PGGT-1 and PGGT-2) (Yokoyama *et al.*, 1998b; Buckner *et al.*, 2000). PFT has recently been more widely studied in parasites. This enzyme catalyzes the transfer of a farnesyl group from farnesyl pyrophosphate to a cysteine SH of a tetrapeptide at the C terminal of the receiving protein. In addition to their function as a protein anchor the prenyl groups can serve as a molecular handle to gain access to different membrane-bound cellular organelles such as Golgi, lysosomes, and mitochondria.

A good example of a prenylated protein in mammals and many lower organisms is the γ subunit of the heterotrimeric G protein that is a component of the cyclic-AMP signal transduction system. Many other smaller GTPases such as members of the Ras protein family involved in mitotic signal transduction are also prenylated. Oncogenic Ras proteins play a role in uncontrolled cell growth and malignant transformation of cells after being farnesylated by PFT. Inhibitors of PFT were shown to reduce Ras farnesylation *in vitro* at micromolar levels. The *in vivo* effects of some of these inhibitors were shown in reduction of tumor size in mice after 2 weeks of treatment. For a comprehensive review see Leonard (1997).

Protein prenylation has also been identified in several protozoal parasites including *T. brucei*, *T. cruzi*, *L. mexicana*, and *P. falciparum*. Concerted research efforts are underway in several laboratories to learn more about prenylation enzymes in these parasites and to identify selective inhibitors. PFT from *T. brucei* has recently been purified and then cloned and expressed in *E. coli*. Inhibitors were identified using a structure that mimics the protein acceptor of the farnesyl group to the cysteine residue at the C-terminal motif, CaaX, where C is a cysteine, a is usually an aliphatic amino acid, and X is a glutamine, serine, or methionine. Peptides where X is Met or Glu have a greater affinity to *T. brucei* PFT. Several inhibitors of the enzyme have been identified. These agents inhibit the growth of the bloodstream stage of *T. brucei* at concentration of 3–10 μM whereas 3T3 fibroblasts were not affected by up to 100 μM. These results are encouraging for using PFT as a target for development of antiparasitic agents (Yokoyama *et al.*, 1998a, 1998b; Buckner *et al.*, 2000).

PFTs have also been studied as targets for antimalarial agents. A group of inhibitors of PFT was chosen on the basis of the CaaX C-terminal sequence recognized by PFTs. Information from the crystal structure of mammalian PFT was used to modify structures of the inhibitors. There was selectivity between the active sites of mammalian and malarial PFT enzymes. A series of nonthiol PFT inhibitors reduced the growth of bloodstream malaria in culture (Ohkanda *et al.*, 2001). It appears that malarial PFT is also a valid drug target for identifying new antimalarials for further testing in animal models.

Cysteine Proteases as Targets for Antitrypanosomal and Antileishmanial Agents

Proteases play an important function in the regulation of the *Trypanosoma* and *Leishmania* life cycle, metabolism, replication, development, and differentiation. For this reason proteases are considered prime targets for choosing selective antiparasitic agents. In-depth knowledge of the structures and functions of these enzymes is essential to facilitate the design of highly selective inhibitors. Although there are a few different types of proteases that have been discovered in trypanosomes and *Leishmania*, the cysteine protease known as cruzain (or cruzipain) is the major one. Cruzain has been the best studied enzyme among protozoal proteases. For reviews see McKerrow *et al.* (1993), Coombs & Mottram (1997), and Rosenthal (1999). The enzyme is encoded by multicopy genes arranged in tandem arrays. The multiplicity of the gene could be to ensure the synthesis of large amounts of the enzyme or to have several isozymes with different functions. The protease has been identified in all the developmental stages of the trypanosomes and different cruzain isozymes may be important for various stages of development and interaction processes with the host. Some functions that have been attributed to cruzain include the invasion of the host cells or allowing the parasite to salvage metabolites from the host. At least two and possibly more variants of cruzain with sequence similarity (84–97% identity) are present in *T. cruzi* during different stages of development and have been studied by different investigators.

Cruzain from *T. cruzi* has been cloned and the amino acid sequence predicted from the DNA sequence is homologous with the enzyme from *T. brucei* as well as with other members of the papain family of proteases (Eakin *et al.*, 1992; Lima *et al.*, 1994). The recombinant enzyme is inhibited by known inhibitors of cysteine proteases, which include leupeptin, N-p-tosyl-L-lysine chloromethyl ketone, and trans-epoxysuccinyl-L-leucylamido(4-guanidino) butane (E-64). The X-ray crystal structure of cruzain complexed with one of the potent irreversible inhibitors, benzyloxycarbonyl-L-phenylalanyl-L-alanine fluoromethyl ketone, was determined at 2.35-Å resolution (McGrath *et al.*, 1995). The presence of this inhibitor at the cruzain active site enhances the resolution of the crystallographic data and provides better information on the binding sites for lead compounds. Although the crystal structure compares to that of papain, there are some distinguishing features that help in identifying selective inhibitors. A new generation of cysteine protease inhibitors have been synthesized to enhance specificity and minimize toxicity. Inhibitors that are fluoromethyl ketone-derivatized pseudopeptides were shown to protect mice from lethal infections of *T. cruzi* without toxicity to the mammalian host (Engel *et al.*, 1998). The results of McKerrow and his colleagues provide the first successful treatment of mice infected with *T. cruzi* with an inhibitor designed to inactivate cruzain.

Structure-based design of cysteine protease inhibitors has been extended to falcipain, a major enzyme in *P. falciparum* trophozoites (see Chapter 4) and to the major cathepsin B-like protease of *Leishmania major*. Both enzymes are members of the papain family and are highly homologous to cruzain (Scheidt *et al.*, 1998). McKerrow's group made use of X-ray crystal structures obtained previously for cruzain (McGrath *et al.*, 1995) to design conformationally constrained inhibitors. A series of pyrrolidinone-containing peptide aldehydes were shown to be highly effective against both the *L. major* cysteine protease and falcipain from *P. falciparum* (Scheidt *et al.*, 1998). Attempts by this group to develop more potent and selective inhibitors are in progress.

In contrast to *T. cruzi*, *Leishmania* species, in addition to having a cysteine protease that is homologous to cruzain, also have at least two other cysteine protease genes: one that is mammalian cathepsin L-like (Mottram *et al.*, 1992) and another that is mammalian cathepsin B-like. The latter was first identified in *L. mexicana* (Bart, Coombs, & Mottram, 1995) and subsequently in *L. major* (Sakanari *et al.*, 1997). These two leishmanial protease genes have been designated cpL (or type II) and cpB (or type III) (Coombs & Mottram, 1997) with cruzain as type I. Although there is strong experimental evidence that these cysteine proteases play a role in macrophage invasion, parasite virulence, and degradation of antibodies, the details of each function have not yet been fully studied.

Two generally known cysteine protease inhibitors, antipain and leupeptin, were reported to be potent inhibitors of *L. mexicana* amastigotes in mouse peritoneal macrophages (Coombs & Baxter, 1984). In the absence of crystallographic analysis of cpL and cpB and in an attempt to speed up the procedure for molecular modeling to screen large numbers of chemicals a new procedure was tried based on similarities among cysteine proteases of known structure (Selzer *et al.*, 1997). The crystal structures of papain, cruzain, and human liver cathepsin B, as well as the limited amino acid sequences obtained for cpL (216 amino acids) and cpB (243 amino acids), were used to compute 3-D structures for the two leishmanial enzymes. Three-dimensional structures of different chemical agents to be tested were also generated. Searching for potential protease inhibitor leads was carried out using the DOCK computer program (Kuntz, 1992; Kuntz, Meng, & Shoichet, 1994). These programs locate possible binding orientations and generate two scores, one for contact and the other for a force field interaction energy. This procedure allowed the investigators to consider 150,000 commercially available compounds as potential inhibitors. Eighteen high-scoring compounds were tested on protease activity (IC_{50} = 50–100 μM) and on cell cultures of *L. major* promastigotes. Of these, three compounds inhibited parasite growth at concentrations of 5–50 μM. None of the compounds had an effect on the growth or appearance of mammalian macrophages in culture. The procedure adopted by Selzer *et al.* (1997) confirms that the search for antiparasitic agents can be facilitated by utilizing computational screens to reduce the time and cost of research.

REFERENCES

Aronov, A. M., Suresh, S., Buckner, F. S., Van Voorhis, W. C., Verlinde, C. L., Opperdoes, F. R., Hol, W. G. & Gelb, M. H. (1999). Structure-based design of submicromolar, biologically active inhibitors of trypanosomatid glyceraldehyde-3-phosphate dehydrogenase. *Proc Natl Acad Sci USA, 96*(8), 4273–4278.

Bacchi, C. J., Nathan, H. C., Hutner, S. H., McCann, P. P. & Sjoerdsma, A. (1980). Polyamine metabolism: A potential therapeutic target in trypanosomes. *Science, 210*(4467), 332–334.

Bacchi, C. J., Garofalo, J., Mockenhaupt, D., McCann, P. P., Diekema, K. A., Pegg, A. E., Nathan, H. C., Mullaney, E. A., Chunosoff, L., Sjoerdsma, A. & Hutner, S. H. (1983). In vivo effects of alpha-DL-difluoromethylornithine on the metabolism and morphology of *Trypanosoma brucei brucei*. *Mol Biochem Parasitol, 7*(3), 209–225.

Bakker, B. M., Westerhoff, H. V., Opperdoes, F. R. & Michels, P. A. (2000). Metabolic control analysis of glycolysis in trypanosomes as an approach to improve selectivity and effectiveness of drugs. *Mol Biochem Parasitol, 106*(1), 1–10.

Barrett, M. P., Tetaud, E., Seyfang, A., Bringaud, F. & Baltz, T. (1998). Trypanosome glucose transporters. *Mol Biochem Parasitol, 91*(1), 195–205.

Bart, G., Coombs, G. H. & Mottram, J. C. (1995). Isolation of *lmcpc*, a gene encoding a *Leishmania mexicana* cathepsin-B-like cysteine proteinase. *Mol Biochem Parasitol, 73*(1–2), 271–274.

Berens, R. L., Krug, E. C. & Marr, J. J. (1995). Purine and pyrimidine metabolism. In J. J. Marr & M. Muller (Eds.), *Biochemistry and Molecular Biology of Parasites* (pp. 89–118). San Diego: Academic Press.

Berman, J. D., Rainey, P. & Santi, D. V. (1983). Metabolism of formycin B by *Leishmania* amastigotes in vitro. Comparative metabolism in infected and uninfected human macrophages. *J Exp Med, 158*(1), 252–257.

Bitonti, A. J., Bacchi, C. J., McCann, P. P. & Sjoerdsma, A. (1985). Catalytic irreversible inhibition of *Trypanosoma brucei brucei* ornithine decarboxylase by substrate and product analogs and their effects on murine trypanosomiasis. *Biochem Pharmacol, 34*(10), 1773–1777.

Bond, C. S., Zhang, Y., Berriman, M., Cunningham, M. L., Fairlamb, A. H. & Hunter, W. N. (1999). Crystal structure of *Trypanosoma cruzi* trypanothione reductase in complex with trypanothione, and the structure-based discovery of new natural product inhibitors. *Structure Fold Des, 7*(1), 81–89.

Borst, P. & Fairlamb, A. H. (1998). Surface receptors and transporters of *Trypanosoma brucei. Annu Rev Microbiol, 52*, 745–778.

Bressi, J. C., Choe, J., Hough, M. T., Buckner, F. S., Van Voorhis, W. C., Verlinde, C. L., Hol, W. G. & Gelb, M. H. (2000). Adenosine analogues as inhibitors of *Trypanosoma brucei* phosphoglycerate kinase: Elucidation of a novel binding mode for a 2-amino-N(6)-substituted adenosine. *J Med Chem, 43*(22), 4135–4150.

Brohn, F. H. & Clarkson, A. B., Jr. (1980). *Trypanosoma brucei brucei*: Patterns of glycolysis at 37 degrees C in vitro. *Mol Biochem Parasitol, 1*(5), 291–305.

Buckner, F. S., Yokoyama, K., Nguyen, L., Grewal, A., Erdjument-Bromage, H., Tempst, P., Strickland, C. L., Xiao, L., Van Voorhis, W. C. & Gelb, M. H. (2000). Cloning, heterologous expression, and distinct substrate specificity of protein farnesyltransferase from *Trypanosoma brucei. J Biol Chem, 275*(29), 21870–21876.

Callahan, H. L. & Beverley, S. M. (1992). A member of the aldoketo reductase family confers methotrexate resistance in *Leishmania. J Biol Chem, 267*(34), 24165–24168.

Carter, N. S. & Fairlamb, A. H. (1993). Arsenical-resistant trypanosomes lack an unusual adenosine transporter. *Nature, 361*(6408), 173–176. [Published erratum appears in *Nature* 1993; *361*(6410), 374.]

Coombs, G. H. & Baxter, J. (1984). Inhibition of *Leishmania* amastigote growth by antipain and leupeptin. *Ann Trop Med Parasitol, 78*(1), 21–24.

Coombs, G. H. & Mottram, J. C. (1997). Parasite proteinases and amino acid metabolism: Possibilities for chemotherapeutic exploitation. *Parasitology, 114*(Suppl), S61–80.

Cunningham, M. L. & Fairlamb, A. H. (1995). Trypanothione reductase from *Leishmania donovani*. Purification, characterisation and inhibition by trivalent antimonials. *Eur J Biochem, 230*(2), 460–468.

Docampo, R. (1990). Sensitivity of parasites to free radical damage by antiparasitic drugs. *Chem Biol Interact, 73*(1), 1–27.

Doering, T. L., Raper, J., Buxbaum, L. U., Adams, S. P., Gordon, J. I., Hart, G. W. & Englund, P. T. (1991). An analog of myristic acid with selective toxicity for African trypanosomes. *Science, 252*(5014), 1851–1854.

Doering, T. L., Lu, T., Werbovetz, K. A., Gokel, G. W., Hart, G. W., Gordon, J. I. & Englund, P. T. (1994). Toxicity of myristic acid analogs toward African trypanosomes. *Proc Natl Acad Sci USA, 91*(21), 9735–9739.

Donald, R. G. K., Carter, D., Ullman, B. & Roos, D. S. (1996). Insertional tagging, cloning, and expression of the *Toxoplasma gondii* hypoxanthine-xanthine-guanine phosphoribosyltransferase gene. Use as a selectable marker for stable transformation. *J Biol Chem, 271*(24), 14010–14019.

Dumas, C., Ouellette, M., Tovar, J., Cunningham, M. L., Fairlamb, A. H., Tamar, S., Olivier, M. & Papadopoulou, B. (1997). Disruption of the trypanothione reductase gene of *Leishmania* decreases its ability to survive oxidative stress in macrophages. *EMBO J, 16*(10), 2590-2598.

Eads, J. C., Scapin, G., Xu, Y., Grubmeyer, C. & Sacchettini, J. C. (1994). The crystal structure of human hypoxanthine-guanine phosphoribosyltransferase with bound GMP. *Cell, 78*(2), 325-334.

Eakin, A. E., Mills, A. A., Harth, G., McKerrow, J. H. & Craik, C. S. (1992). The sequence, organization, and expression of the major cysteine protease (cruzain) from *Trypanosoma cruzi. J Biol Chem, 267*(11), 7411-7420.

Eisenthal, R., Game, S. & Holman, G. D. (1989). Specificity and kinetics of hexose transport in *Trypanosoma brucei. Biochim Biophys Acta, 985*(1), 81-89.

el Kouni, M. H., Naguib, F. N., Panzica, R. P., Otter, B. A., Chu, S. H., Gosselin, G., Chu, C. K., Schinazi, R. F., Shealy, Y. F., Goudgaon, N., Ozerov, A. A., Ueda, T. & Iltzsch, M. H. (1996). Effects of modifications in the pentose moiety and conformational changes on the binding of nucleoside ligands to uridine phosphorylase from *Toxoplasma gondii. Biochem Pharmacol, 51*(12), 1687-1700.

Engel, J. C., Doyle, P. S., Hsieh, I. & McKerrow, J. H. (1998). Cysteine protease inhibitors cure an experimental *Trypanosoma cruzi* infection. *J Exp Med, 188*(4), 725-734.

Englund, P., Ferguson, D., Guilbride, D., Johnson, C., Li, C., Perez-Morga, D., Rocco, L. & Torri, A. (1995). The replication of kinetoplast DNA. In J. Boothroyd & R. Komuniecki (Eds.), *Molecular Approaches to Parasitology* (Vol. 12, pp. 147-161). New York: Wiley-Liss.

Fairlamb, A. (1982). Biochemistry of trypanosomiasis and rational approaches to chemotherapy. *Trends Biochem Sci, 7*, 249-253.

Fairlamb, A. H. & Bowman, I. B. (1980). Uptake of the trypanocidal drug suramin by bloodstream forms of *Trypanosoma brucei* and its effect on respiration and growth rate in vivo. *Mol Biochem Parasitol, 1*(6), 315-333.

Fairlamb, A. H. & Cerami, A. (1992). Metabolism and functions of trypanothione in the Kinetoplastida. *Annu Rev Microbiol, 46*, 695-729.

Fairlamb, A. H., Henderson, G. B. & Cerami, A. (1989). Trypanothione is the primary target for arsenical drugs against African trypanosomes. *Proc Natl Acad Sci USA, 86*(8), 2607-2611.

Ferguson, M. A. & Cross, G. A. (1984). Myristylation of the membrane form of a *Trypanosoma brucei* variant surface glycoprotein. *J Biol Chem, 259*(5), 3011-3015.

Focia, P. J., Craig, S. P., III, Nieves-Alicea, R., Fletterick, R. J. & Eakin, A. E. (1998). A 1.4 A crystal structure for the hypoxanthine phosphoribosyltransferase of *Trypanosoma cruzi. Biochemistry, 37*(43), 15066-15075.

Garforth, J., Yin, H., McKie, J. H., Douglas, K. T. & Fairlamb, A. H. (1997). Rational design of selective ligands for trypanothione reductase from *Trypanosoma cruzi*. Structural effects on the inhibition by dibenzazepines based on imipramine. *J Enzyme Inhib, 12*(3), 161-173.

Goodwin, L. & Page, J. (1943). A study of the excretion of organic antimonials using a polarographic procedure. *Biochem J, 37*, 189-209.

Gourley, D. G., Luba, J., Hardy, L. W., Beverley, S. M. & Hunter, W. N. (1999). Crystallization of recombinant *Leishmania major* pteridine reductase 1 (PTR1). *Acta Crystallogr D Biol Crystallogr, 55*(Pt 9), 1608-1610.

Hardy, L. W., Matthews, W., Nare, B. & Beverley, S. M. (1997). Biochemical and genetic tests for inhibitors of *Leishmania* pteridine pathways. *Exp Parasitol, 87*(3), 157–169.

Hofmann, B., Budde, H., Bruns, K., Guerrero, S. A., Kalisz, H. M., Menge, U., Montemartini, M., Nogoceke, E., Steinert, P., Wissing, J. B., Flohe, L. & Hecht, H. J. (2001). Structures of tryparedoxins revealing interaction with trypanothione. *Biol Chem, 382*(3), 459–471.

Hunter, W. N., Bailey, S., Habash, J., Harrop, S. J., Helliwell, J. R., Aboagye-Kwarteng, T., Smith, K. & Fairlamb, A. H. (1992). Active site of trypanothione reductase. A target for rational drug design. *J Mol Biol, 227*(1), 322–333.

Iltzsch, M. H. (1993). Pyrimidine salvage pathways in *Toxoplasma gondii*. *J Eukaryot Microbiol, 40*(1), 24–28.

Iltzsch, M. H., Uber, S. S., Tankersley, K. O. & el Kouni, M. H. (1995). Structure–activity relationship for the binding of nucleoside ligands to adenosine kinase from *Toxoplasma gondii*. *Biochem Pharmacol, 49*(10), 1501–1512.

Iten, M., Mett, H., Evans, A., Enyaru, J. C., Brun, R. & Kaminsky, R. (1997). Alterations in ornithine decarboxylase characteristics account for tolerance of *Trypanosoma brucei* rhodesiense to D,L-alpha-difluoromethylornithine. *Antimicrob Agents Chemother, 41*(9), 1922–1925.

Jardim, A., Bergeson, S. E., Shih, S., Carter, N., Lucas, R. W., Merlin, G., Myler, P. J., Stuart, K. & Ullman, B. (1999). Xanthine phosphoribosyltransferase from *Leishmania donovani*. Molecular cloning, biochemical characterization, and genetic analysis. *J Biol Chem, 274*(48), 34403–34410.

Krieger, S., Schwarz, W., Ariyanayagam, M. R., Fairlamb, A. H., Krauth-Siegel, R. L. & Clayton, C. (2000). Trypanosomes lacking trypanothione reductase are avirulent and show increased sensitivity to oxidative stress. *Mol Microbiol, 35*(3), 542–552.

Krug, E. C., Marr, J. J. & Berens, R. L. (1989). Purine metabolism in *Toxoplasma gondii*. *J Biol Chem, 264*(18), 10601–10607.

Kuntz, I., Meng, E. & Shoichet, B. (1994). Structure-based molecular design. *Accounts Chem Res, 27,* 117–123.

Kuntz, I. D. (1992). Structure-based strategies for drug design and discovery. *Science, 257*(5073), 1078–1082 [See comments.]

Leonard, D. M. (1997). Ras farnesyltransferase: A new therapeutic target. *J Med Chem, 40*(19), 2971–2990.

Li, F., Hua, S. B., Wang, C. C. & Gottesdiener, K. M. (1996). Procyclic *Trypanosoma brucei* cell lines deficient in ornithine decarboxylase activity. *Mol Biochem Parasitol, 78*(1–2), 227–236.

Lima, A. P., Tessier, D. C., Thomas, D. Y., Scharfstein, J., Storer, A. C. & Vernet, T. (1994). Identification of new cysteine protease gene isoforms in *Trypanosoma cruzi*. *Mol Biochem Parasitol, 67*(2), 333–338.

Marr, J. J. (1991). Purine analogs as chemotherapeutic agents in leishmaniasis and American trypanosomiasis. *J Lab Clin Med, 118*(2), 111–119.

Marr, J. J. & Ullman, B. (1995). Concepts of chemotherapy. In J. J. Marr & M. Muller (Eds.), *Biochemistry and Molecular Biology of Parasites* (pp. 323–336). San Diego: Academic Press.

Martinez, S. & Marr, J. J. (1992). Allopurinol in the treatment of American cutaneous leishmaniasis. *N Engl J Med, 326*(11), 741–744 [See comments.]

McGrath, M. E., Eakin, A. E., Engel, J. C., McKerrow, J. H., Craik, C. S. & Fletterick, R. J. (1995). The crystal structure of cruzain: A therapeutic target for Chagas' disease. *J Mol Biol, 247*(2), 251–259.

McKerrow, J. H., Sun, E., Rosenthal, P. J. & Bouvier, J. (1993). The proteases and pathogenicity of parasitic protozoa. *Annu Rev Microbiol, 47*, 821–853.

Miles, R. W., Tyler, P. C., Evans, G. B., Furneaux, R. H., Parkin, D. W. & Schramm, V. L. (1999). Iminoribitol transition state analogue inhibitors of protozoan nucleoside hydrolases. *Biochemistry, 38*(40), 13147–13154.

Milord, F., Pepin, J., Loko, L., Ethier, L. & Mpia, B. (1992). Efficacy and toxicity of eflornithine for treatment of *Trypanosoma brucei gambiense* sleeping sickness. *Lancet, 340*(8820), 652–655.

Morita, Y. S., Paul, K. S. & Englund, P. T. (2000). Specialized fatty acid synthesis in African trypanosomes: Myristate for GPI anchors. *Science, 288*(5463), 140–143.

Mottram, J. C., Robertson, C. D., Coombs, G. H. & Barry, J. D. (1992). A developmentally regulated cysteine proteinase gene of *Leishmania mexicana. Mol Microbiol, 6*(14), 1925–1932.

Nare, B., Luba, J., Hardy, L. W. & Beverley, S. (1997). New approaches to *Leishmania* chemotherapy: Pteridine reductase 1 (PTR1) as a target and modulator of antifolate sensitivity. *Parasitology, 114*(Suppl), S101–110.

Nathan, H. C., Bacchi, C. J., Hutner, S. H., Rescigno, D., McCann, P. P. & Sjoerdsma, A. (1981). Antagonism by polyamines of the curative effects of alpha-difluoromethylornithine in *Trypanosoma brucei brucei* infections. *Biochem Pharmacol, 30*(21), 3010–3013.

Nelson, D. L. & Cox, M. M. (2000). *Leninger Principles of Biochemistry* (3rd ed.) pp. 845–847. New York: Worth.

Ohkanda, J., Lockman, J. W., Yokoyama, K., Gelb, M. H., Croft, S. L., Kendrick, H., Harrell, M. I., Feagin, J. E., Blaskovich, M. A., Sebti, S. M. & Hamilton, A. D. (2001). Peptidomimetic inhibitors of protein farnesyltransferase show potent antimalarial activity. *Bioorg Med Chem Lett, 11*(6), 761–764.

Paul, K., Jiang, D., Morita, Y. & Englund, P. (2001). Fatty acid synthesis in Aftican trypanosomes: A solution to the myristate mystery. *Trends Parasitol, 17*(8), 381–387.

Pelle, R., Schramm, V. L. & Parkin, D. W. (1998). Molecular cloning and expression of a purine-specific N-ribohydrolase from *Trypanosoma brucei brucei*. Sequence, expression, and molecular analysis. *J Biol Chem, 273*(4), 2118–2126.

Phillips, M. A., Coffino, P. & Wang, C. C. (1987). Cloning and sequencing of the ornithine decarboxylase gene from *Trypanosoma brucei*. Implications for enzyme turnover and selective difluoromethylornithine inhibition. *J Biol Chem, 262*(18), 8721–8727.

Plagemann, P. G., Wohlhueter, R. M. & Woffendin, C. (1988). Nucleoside and nucleobase transport in animal cells. *Biochim Biophys Acta, 947*(3), 405–443.

Poulin, R., Lu, L., Ackermann, B., Bey, P. & Pegg, A. E. (1992). Mechanism of the irreversible inactivation of mouse ornithine decarboxylase by alpha-difluoromethylornithine. Characterization of sequences at the inhibitor and coenzyme binding sites. *J Biol Chem, 267*(1), 150–158.

Queen, S. A., Vander Jagt, D. & Reyes, P. (1988). Properties and substrate specificity of a purine phosphoribosyltransferase from the human malaria parasite, *Plasmodium falciparum. Mol Biochem Parasitol, 30*(2), 123–133.

Rainey, P. & Santi, D. V. (1983). Metabolism and mechanism of action of formycin B in *Leishmania. Proc Natl Acad Sci USA, 80*(1), 288–292.

Robello, C., Navarro, P., Castanys, S. & Gamarro, F. (1997). A pteridine reductase gene ptr1 contiguous to a P-glycoprotein confers resistance to antifolates in *Trypanosoma cruzi*. *Mol Biochem Parasitol*, *90*(2), 525–535.

Rogers, S., Wells, R. & Rechsteiner, M. (1986). Amino acid sequences common to rapidly degraded proteins: The PEST hypothesis. *Science*, *234*(4774), 364–368.

Rosenthal, P. J. (1999). Proteases of protozoan parasites. *Adv Parasitol*, *43*, 105–159.

Sakanari, J. A., Nadler, S. A., Chan, V. J., Engel, J. C., Leptak, C. & Bouvier, J. (1997). *Leishmania major*: Comparison of the cathepsin L-and B-like cysteine protease genes with those of other trypanosomatids. *Exp Parasitol*, *85*(1), 63–76.

Scheidt, K. A., Roush, W. R., McKerrow, J. H., Selzer, P. M., Hansell, E. & Rosenthal, P. J. (1998). Structure-based design, synthesis and evaluation of conformationally constrained cysteine protease inhibitors. *Bioorg Med Chem*, *6*(12), 2477–2494.

Schwab, J. C., Afifi Afifi, M., Pizzorno, G., Handschumacher, R. E. & Joiner, K. A. (1995). *Toxoplasma gondii* tachyzoites possess an unusual plasma membrane adenosine transporter. *Mol Biochem Parasitol*, *70*(1–2), 59–69.

Selzer, P. M., Chen, X., Chan, V. J., Cheng, M., Kenyon, G. L., Kuntz, I. D., Sakanari, J. A., Cohen, F. E. & McKerrow, J. H. (1997). *Leishmania major*: Molecular modeling of cysteine proteases and prediction of new nonpeptide inhibitors. *Exp Parasitol*, *87*(3), 212–221.

Seyfang, A. & Duszenko, M. (1991). Specificity of glucose transport in *Trypanosoma brucei*. Effective inhibition by phloretin and cytochalasin B. *Eur J Biochem*, *202*(1), 191–196.

Shapiro, T. A. (1994). Drugs affecting trypanosome topoisomerases. *Adv Pharmacol*, *29*(B), 187–200.

Shapiro, T. A. & Englund, P. T. (1990). Selective cleavage of kinetoplast DNA minicircles promoted by antitrypanosomal drugs. *Proc Natl Acad Sci USA*, *87*(3), 950–954.

Sjoerdsma, A. & Schechter, P. J. (1984). Chemotherapeutic implications of polyamine biosynthesis inhibition. *Clin Pharmacol Ther*, *35*(3), 287–300.

Smith, K., Nadeau, K., Bradley, M., Walsh, C. & Fairlamb, A. H. (1992). Purification of glutathionylspermidine and trypanothione synthetases from *Crithidia fasciculata*. *Protein Sci*, *1*(7), 874–883.

Smyth, J. D. (1994). *Introduction to Animal Parasitology* (3rd ed.), pp. 58–87. Cambridge: Cambridge University Press.

Sullivan, W. J., Jr., Chiang, C. W., Wilson, C. M., Naguib, F. N., el Kouni, M. H., Donald, R. G. & Roos, D. S. (1999). Insertional tagging of at least two loci associated with resistance to adenine arabinoside in *Toxoplasma gondii*, and cloning of the adenosine kinase locus. *Mol Biochem Parasitol*, *103*(1), 1–14.

Suresh, S., Bressi, J. C., Kennedy, K. J., Verlinde, C. L., Gelb, M. H. & Hol, W. G. (2001). Conformational changes in *Leishmania mexicana* glyceraldehyde-3-phosphate dehydrogenase induced by designed inhibitors. *J Mol Biol*, *309*(2), 423–435.

Tabor, C. W. & Tabor, H. (1984). Polyamines. *Annu Rev Biochem*, *53*, 749–790.

Tuttle, J. V. & Krenitsky, T. A. (1980). Purine phosphoribosyltransferases from *Leishmania donovani*. *J Biol Chem*, *255*(3), 909–916.

Ullman, B. & Carter, D. (1995). Hypoxanthine-guanine phosphoribosyltransferase as a therapeutic target in protozoal infections. *Infect Agents Dis*, *4*(1), 29–40.

Ullman, B. & Carter, D. (1997). Molecular and biochemical studies on the hypoxanthine-guanine phosphoribosyltransferases of the pathogenic haemoflagellates. *Int J Parasitol*, *27*(2), 203–213.

Van Hellemond, J. J., Opperdoes, F. R. & Tielens, A. G. (1998). Trypanosomatidae produce acetate via a mitochondrial acetate:succinate CoA transferase. *Proc Natl Acad Sci USA, 95*(6), 3036–3041.

Visser, N., Opperdoes, F. R. & Borst, P. (1981). Subcellular compartmentation of glycolytic intermediates in *Trypanosoma brucei*. *Eur J Biochem, 118*(3), 521–526.

Waller, R. F., Keeling, P. J., Donald, R. G., Striepen, B., Handman, E., Lang-Unnasch, N., Cowman, A. F., Besra, G. S., Roos, D. S. & McFadden, G. I. (1998). Nuclear-encoded proteins target to the plastid in *Toxoplasma gondii* and *Plasmodium falciparum*. *Proc Natl Acad Sci USA, 95*(21), 12352–12357.

Wang, C. C. (1995). Molecular mechanisms and therapeutic approaches to the treatment of African trypanosomiasis. *Annu Rev Pharmacol Toxicol, 35*, 93–127.

Werbovetz, K. A., Bacchi, C. J. & Englund, P. T. (1996). Trypanocidal analogs of myristate and myristoyllysophosphatidylcholine. *Mol Biochem Parasitol, 81*(1), 115–118.

Wiemer, E. A., Michels, P. A. & Opperdoes, F. R. (1995). The inhibition of pyruvate transport across the plasma membrane of the bloodstream form of *Trypanosoma brucei* and its metabolic implications. *Biochem J, 312*(Pt 2), 479–484.

Wiemer, E. A., Ter Kuile, B. H., Michels, P. A. & Opperdoes, F. R. (1992). Pyruvate transport across the plasma membrane of the bloodstream form of *Trypanosoma brucei* is mediated by a facilitated diffusion carrier. *Biochem Biophys Res Commun, 184*(2), 1028–1034.

Willson, M., Callens, M., Kuntz, D. A., Perie, J. & Opperdoes, F. R. (1993). Synthesis and activity of inhibitors highly specific for the glycolytic enzymes from *Trypanosoma brucei*. *Mol Biochem Parasitol, 59*(2), 201–210.

Yokoyama, K., Trobridge, P., Buckner, F. S., Scholten, J., Stuart, K. D., Van Voorhis, W. C. & Gelb, M. H. (1998a). The effects of protein farnesyltransferase inhibitors on trypanosomatids: Inhibition of protein farnesylation and cell growth. *Mol Biochem Parasitol, 94*(1), 87–97.

Yokoyama, K., Trobridge, P., Buckner, F. S., Van Voorhis, W. C., Stuart, K. D. & Gelb, M. H. (1998b). Protein farnesyltransferase from *Trypanosoma brucei*. A heterodimer of 61- and 65-kDa subunits as a new target for antiparasite therapeutics. *J Biol Chem, 273*(41), 26497–26505.

Zhang, Y., Bailey, S., Naismith, J. H., Bond, C. S., Habash, J., McLaughlin, P., Papiz, M. Z., Borges, A., Cunningham, M., Fairlamb, A. H. *et al.* (1993). *Trypanosoma cruzi* trypanothione reductase. Crystallization, unit cell dimensions and structure solution. *J Mol Biol, 232*(4), 1217–1220.

Zhang, Y., Bond, C. S., Bailey, S., Cunningham, M. L., Fairlamb, A. H. & Hunter, W. N. (1996). The crystal structure of trypanothione reductase from the human pathogen *Trypanosoma cruzi* at 2.3 A resolution. *Protein Sci, 5*(1), 52–61.

6

Targets in Amitochondrial Protists

Biology of Amitochondrial Protists

The parasitic protozoa discussed in this chapter are grouped together on the basis of the nature of their fermentative energy metabolism. In addition to having no mitochondria, they undergo no cytochrome-mediated electron transport and no oxidative phosphorylation processes. These organisms, which do not live intracellularly, have metabolically adapted to survive under strictly anaerobic conditions or under low levels of oxygen tension in the lumen of the gut or in the vagina of their hosts. They are represented here by *Entamoeba histolytica*, *Giardia intestinalis*, and *Trichomonas vaginalis*. All these parasites are aerotolerant and take up O_2 that is present in their natural habitat. The O_2 does not appear to be involved in energy production but is used as a means of detoxifying potentially noxious or toxic materials in their environment (Coombs & Muller, 1995). Looking at them from the evolutionary point of view they appear to have evolved at a time when life was predominantly anaerobic and mitochondria were unknown. Because of this and other metabolic differences there is a view that amitochondrial eukaryotes have separated early from the main trunk of eukaryotic evolution (Muller, 1988). However, a report by Mertens *et al.* (1998) on the marked diversity of pyrophosphate-dependent PFK sequences of some protists makes this conclusion less certain. These organisms have pyruvate:ferredoxin oxidoreductases that function as electron transport proteins. Others, especially *Trichomonas spp.*, have developed new organelles called hydrogenosomes, which are compartments for energy metabolism. Under anaerobic conditions these organelles produce molecular hydrogen by oxidizing pyruvate or malate (Muller, 1993; Martin & Muller, 1998).

Entamoeba histolytica

Among these parasitic protists *Entamoeba histolytica*, the causative organism of intestinal amebiasis in humans, is the most important clinically. It is found in arctic, temperate, and tropical climates worldwide and is reported to affect 480 million people. The term "histolytica" refers to the parasite's habit of "tissue dissolving." They are actually scavengers feeding mainly on bacteria, tissue fragments, starch, and other nutrients in the host's intestinal contents. One factor that slowed research on these parasites is the difficulty of cultivating them in the laboratory. For a while it was possible to maintain cultures only in association with bacteria or other protozoa. This made it difficult to assign biological information obtained from these mixed cultures to any specific organism. It was not until the late 1970s that axenic cultures were developed for *E. histolytica* and other related amebic parasites (Diamond *et al.*, 1978). Most of the biological studies have been done on the cultured trophozoites and very little information is available on the cyst stage.

The ameba parasite has three distinct stages in the life cycle: trophozoite, precyst, and cyst. Humans become infected by ingesting mature cysts from fecally contaminated food, water, or fingers. After several divisions in the lower bowel, the next developmental stage, the trophozoites, emerge. The uninucleated trophozoites, which use pseudopods for motility, are the actively growing form of the parasite and cause the clinical symptoms. The trophozoites move to the cecal area of the large intestine, which is the optimal site for colonization, but they may metastasize to the liver and other extraintestinal sites (lung and brain). The cyst stage is formed if the bowel passage of the protozoan is not rapid. This is the stage that is transmittable to the next host.

Giardia intestinalis (G. duodenalis)

Giardia intestinalis, commonly called *G. lamblia*, is a flagellate protozoa that inhabits the lumen of the small intestine, causing giardiasis, a severe diarrheal disease. Formation of cysts occurs in the proximal part of the small intestine. Giardiasis is transmitted via ingestion of the parasite cysts with food or water. The intensity of infection varies from no symptoms (asymptomatic) to severe diarrhea frequently associated with nausea and abdominal discomfort. It is found worldwide with high prevalence in developing countries. In the United States it is found principally in mountain streams in parks and wilderness areas, and occasionally in urban settings.

Trichomonas vaginalis

Infection by *Trichomonas vaginalis* (the causative agent of trichomoniasis) is highly specific to the genitourinary tract. In the United States it accounts for 3 million infections per year. The site of infection in women is mainly the vagina and the urethra, where it could cause severe inflammation and vaginal discharge. Transmission is through sexual contact. Infection in men is almost asymptomatic but could cause predisposition to nongonococcal urethritis (Strickland, 1991).

Carbohydrate Metabolism of *Entamoeba histolytica*

In the 1950s a movement arose among some scientists in the United States for a rational approach to the chemotherapy of amebic dysentery in man. When axenic culturing of the parasite in the laboratory was established, it was discovered that both exogenous glucose and endogenous glycogen are metabolized by *E. histolytica* with ethanol, acetate, and CO_2 as the main end products. *E. histolytica* is not strictly anaerobic because it can survive in and can utilize O_2 if it is available. The proportions of the metabolic products vary depending on the degree of anaerobiasis. Glucose taken up by the parasite is stored as glycogen or metabolized through the Embden–Meyerhof pathway only as far as pyruvate because the parasite lacks lactic dehydrogenase. Thereafter, it has some unique and interesting differences. The glycolytic pathway has been reviewed by McLaughlin & Aley (1985) and Mertens (1993). The main glycolytic reactions beyond phosphoenolpyruvate are shown in Fig. 6.1.

A distinctive aspect of glycolysis in *E. histolytica* is the parasite's use of pyrophosphate (PP_i) rather than ATP as an energy phosphoryl donor for several glycolytic reactions. This substitution is an ingenious way for the parasite to save ATP. PP_i is a product of many catabolic reactions in DNA and glycogen synthesis. PP_i has almost as much energy as the gamma phosphate in ATP. The mammalian hosts do not use PP_i as an energy source because they have very active pyrophosphatases that waste this energy by hydrolyzing PP_i to P_i. The demonstrated lack of cytosolic pyrophosphatases in *E. histolytica* corresponds well with the ability of these parasites to use PP_i as an energy source (McLaughlin, Lindmark, & Muller, 1978). *E. histolytica* does not have fructose bisphosphatase, the enzyme that reverses the PFK reaction. Unlike ATP-PFK in mammals PP_i-PFK in these parasites is a reversible enzyme and functions as a key enzyme in gluconeogenesis. PP_i-PFK of *E. histolytica* has been shown to have a latent nucleotide binding site, suggesting that ATP-PFK is the more primitive

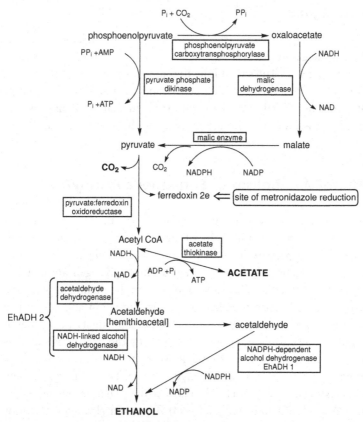

Figure 6.1. Main enzymatic steps in the metabolism of phosphoenolpyruvate to pyruvate and its oxidative decarboxylation to acetylCoA in *Entamoeba histolytica*. Subsequent reactions show the production of acetate and ethanol (metabolic end products). The diagram was adapted from figures in a review by McLaughlin & Aley (1985).

enzyme (Chi & Kemp, 2000). PP_i-PFK has a million-fold preference for PP_i over ATP. The introduction of a single mutation by site-directed mutagenesis changes the affinity of PFK from PP_i to ATP.

The following enzyme steps in *E. histolytica* are considered to be quite different from homologous reactions in the host and have been reviewed by Reeves, a pioneer in this field, who, with his colleagues, has contributed a great deal to our understanding of the unusual metabolic pathways in this parasite (Reeves, 1984).

In addition to phosphofructokinase, another enzyme, pyruvate phosphate dikinase (PPDK), uses PP_i as a phosphoryl donor in converting phosphoenolpyruvate to pyruvate. This is quite different from the analogous enzyme in the host, the pyruvic kinase that is part of the Embden–Meyerhof scheme. PPDK catalyzes the reversible transfer of the phosphate group from phosphoenol-pyruvate and one phosphate from PP_i to the β and γ positions of AMP, respectively, forming one molecule of ATP (Reeves, Menzies, & Hsu, 1968).

An alternative way to form pyruvate from phosphoenolpyruvate utilizes three reactions (Fig. 6.1). First, there is a CO_2-fixing step catalyzed by phospho-enolpyruvate carboxytransphosphorylase forming oxaloacetate. This is also a PP_i enzyme that catalyzes a reaction analogous to that of PEP carboxykinase (GDP/GTP linked) in mammals. The next reaction is the NADH-dependent malic dehydrogenase that reduces oxaloacetate to malate and then the malic enzyme that decarboxylates malate to pyruvate accompanied by reduction of NADP to NADPH.

E. histolytica has a pyruvate:ferredoxin oxidoreductase (PFOR) that catalyzes the transformation of pyruvate and Coenzyme A (CoA) to acetyl CoA, CO_2, and two electrons, which are transferred to ferredoxin. Ferredoxin is a low molec-ular weight protein that contains iron–sulfur 2(4Fe–4S) clusters and functions as an electron acceptor. The electrons are then transferred to cofactors of relatively low redox such as NAD or NADP. The precise mechanism of reoxidation of ferredoxin in *E. histolytica* and *Giardia* is not yet known but may involve pro-viding reducing equivalents for ethanol production (Muller, 1988). Acetyl CoA can be metabolized to acetate by the enzyme acetate thiokinase using ADP and P_i. In addition to acetate one mole of ATP is formed in this reaction. Reeves and Guthrie discovered that *E. histolytica* has another acetate kinase that uses PP_i as a phosphoryl donor to form acetyl phosphate. This is different from the acetate thiokinase. The enzyme has been partially purified and it was shown that ATP cannot replace PP_i as a phosphoryl donor. Its role in the energy metabolism of the parasite has not been determined (Reeves & Guthrie, 1975).

Acetyl CoA can be reduced to acetaldehyde, which is conjugated as a thio-hemiacetal and finally reduced to ethanol. There are two types of alcohol dehy-drogenases identified in *E. histolytica*. Both enzymes have been cloned and their substrates have been identified. *E. histolytica* alcohol dehydrogenase 1 (EhADH1) is NADP dependent whereas the second alcohol dehydrogenase (EhADH2) is NAD dependent. Both enzymes can reduce acetaldehyde as well as ethanol as substrates (Lo & Reeves, 1978; Reeves, 1984; Kumar *et al.*, 1992; Yang, Kairong, & Stanley, 1994). EhADH2, the NAD-dependent enzyme, is believed to be of major importance in alcohol metabolism in *E. histolytica* but the role of the NADP-dependent alcohol dehydrogenase (EhADH1) has not been determined. EhADH2 was purified and characterized (Bruchhaus & Tannich, 1994). The re-sults established two catalytic functions for the enzyme: an NAD^+-dependent acetaldehyde and an alcohol dehydrogenase activity in the same molecule. Kinetic studies on this bifunctional enzyme revealed a lower K_m for acetyl CoA (0.015 mM) than for acetaldehyde (0.15 mM) and a relatively high K_m of 80 mM for ethanol, which indicates that EhADH2 functions in the direction of ethanol formation. The identification of the bifunctional nature of EhADH2 introduces it as a unique target that is not present in the host but is essential to the energy metabolism of the parasite. It is a good candidate to use for selection of new antiamebic agents. The amount of oxygen available to the organism determines the percentage of metabolites utilizing these alternative pathways. Under aerobic conditions acetate, CO_2, and ethanol are formed, whereas under anaerobic conditions only CO_2 and ethanol are produced by the organism (Muller, 1988).

Energy Metabolism of *Giardia intestinalis*

G. intestinalis metabolizes glucose via the same glycolytic enzymes used by *E. histolytica* (for a review see Mertens (1993)). These enzymes include a PP_i-PFK and PPDK instead of ATP-PFK and pyruvic kinase. PFOR coupled to ferredoxin is also used for the oxidative decarboxylation of pyruvate. The trophozoites of *G. intestinalis* adhere to the mucosal lining of the epithelial cells of the duodenum where conditions vary from anaerobic to low levels of oxygen tension. Production of metabolic products, which include acetate, ethanol, and CO_2, are influenced by changes in the O_2 tension of the trophozoite environment. The rate of CO_2 and acetate production is increased whereas the ethanol production rate is reduced slightly under anaerobic conditions.

In addition to their need for glucose to grow, *G. intestinalis* uses arginine as a major source of energy and for ATP formation (Edwards *et al.*, 1992). Arginine was particularly needed during the initial stages of rapid growth. Even in medium with no glucose, addition of arginine to the medium is sufficient to increase cell growth during the first two days of the culture. Furthermore, depletion of arginine from the culture medium leads to cessation of trophozoite growth. The metabolism of arginine with concomitant production of ATP in the parasites is carried out via the arginine dihydrolase pathway, which was previously thought to be restricted to prokaryotes. The three enzymes in this pathway are all present in *G. intestinalis*. They are arginine deiminase (6-arginine iminohydrolase), ornithine transcarbamoylase (carbamoyl phosphate:L-ornithine carbamoyl transferase), and carbamate kinase (ATP:carbamate phosphotransferase) (Edwards *et al.*, 1992; Schofield *et al.*, 1992). The sum of the arginine dihydrolase pathway reactions for energy production is

$$\text{arginine} + \text{ADP} + P_i + H_2O \rightarrow \text{ornithine} + 2NH_3 + CO_2 + \text{ATP}.$$

The carbamate kinase gene from *G. intestinalis* has been cloned, expressed in *E. coli*, and characterized. Both the native and the purified recombinant enzyme have similar physical properties (Minotto *et al.*, 1999).

In addition to acetate, ethanol, and CO_2, alanine was identified by proton NMR spectroscopy as a major metabolic product of *Giardia* under anaerobic conditions. (Edwards *et al.*, 1989; Paget *et al.*, 1990). Aerobically, the production of both alanine and ethanol are inhibited and more acetate and CO_2 are formed. NMR spectroscopy using ^{13}C-labeled glucose showed that alanine is derived from glucose metabolism. The metabolic pathway for the production of alanine has been postulated to include direct formation via L-alanine dehydrogenase from pyruvate and ammonia (a metabolic product of the arginine dihydrolase pathway).

These findings showing arginine as an essential energy source for *Giardia* and alanine as a metabolic product draws our attention to an area of parasite biochemistry that has been neglected: the metabolism of amino acids in parasites.

Hydrogenosomes and Energy Metabolism in Trichomonads

The trichomonads are a group of protists that have a membrane-bound organelle called the hydrogenosome. Although these organelles are occasionally compared with mitochondria because both structures have energy-producing enzymes, the hydrogenosomes are distinctly different. They contain the basic enzymes for oxidative decarboxylation of pyruvate, which is coupled to ferredoxin, the iron–sulfur electron carrier protein. Hydrogenosomes also have an enzyme system that transfers electrons from ferredoxin to molecular hydrogen. Muller (1993) wrote a comprehensive review on this subject. In addition to trichomonads, several other free-living protist groups have been reported to have organelles with similar structure and function. All these have one feature in common: They live in an environment that is either anaerobic or has low oxygen tension. The presence of hydrogenosomes in different protists that are located far from each other on the evolutionary eukaryotic tree has encouraged evolutionary biologists to hypothesize about their ancestry and how they have evolved. Because of structural similarities between the hydrogenosomes and mitochondria a case has been made that the hydrogenosomes were derived from mitochondria (Finlay & Fenchel, 1989). There is also some support for the endosymbiotic hypothesis through DNA-containing organelles from anaerobic prokaryotes rather than the evolutionary conversion from mitochondria (Muller, 1993). *Clostridia* has a similar pyruvate metabolism. More genomic information on hydrogenosomes of a variety of anaerobic protozoa belonging to different genera may be needed to ascertain their origin. Hydrogenosomes are not present in either *Entamoeba* or *Giardia*.

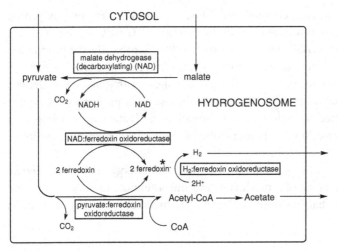

Figure 6.2. Some of the pathways in the hydrogenosome for the metabolism of pyruvate to produce CO_2, acetate and hydrogen. Adapted from Muller (1993) with permission. (★) indicates where metronidazole (if present) takes two electrons from reduced ferredoxin and becomes cytotoxic.

Trichomonas vaginalis and *Tritrichomonas foetus* (a parasite of the genital tract of cattle) are two representative amitochondrial protists that have hydrogenosomes. Both parasites, like *E. histolytica* and *Giardia*, use PP_i-PFK instead of ATP-PFK. However, these trichomonads do not have pyruvate phosphate dikinase, which is present in *E. histolytica* and *Giardia*. Instead, pyruvic kinase is present in *T. vaginalis* and converts PEP to pyruvate and phosphorylates ADP to ATP. Pyruvic kinase from *T. vaginalis* has been purified and characterized (Mertens, Van Schaftingen, & Muller, 1992). The enzyme appears to have allosteric kinetics that are different from other pyruvic kinases including those from mammals. The *T. vaginalis* pyruvic kinase exhibits positive allosteric cooperative kinetics for phosphoenol pyruvate without affecting maximal activity. It is, however, unaffected by known activators such as fructose 1,6-bisphosphate and fructose 2,6-bisphosphate. Instead, it is activated by ribose 5-P and by glycerate 3-P, two intermediary metabolites not previously recognized to affect eukaryotic pyruvic kinase.

T. foetus has no pyruvate dikinase nor detectable pyruvic kinase activity. Phosphoenol pyruvate appears to be converted to pyruvate in the cytosol via PEP carboxykinase, NAD-malate dehydrogenase, and NADP-malic enzyme (see Fig. 6.1) (Hrdy, Mertens, & Van Schaftingen, 1993).

Under anaerobic conditions pyruvate produced from glycolysis is taken up by the hydrogenosomes (Muller, 1993; Fig. 6.2). In the hydrogenosome pyruvate is oxidatively decarboxylated, producing CO_2 and acetate that forms (with

CoA) acetyl CoA. This reaction is catalyzed by PFOR. Ferredoxin accepts two electrons, which are transferred by H_2:ferredoxin oxidoreductase to H_2, which subsequently passes out of the hydrogenosome into the cytosol of the cell. Under anaerobic conditions equimolar amounts of CO_2, H_2, and acetate are produced from pyruvate in isolated intact hydrogenosomes of *T. vaginalis*. In addition to the pyruvate taken into the hydrogenosome, some pyruvate may be produced in the hydrogenosome by malate dehydrogenase (decarboxylating) (NAD) and electrons may be transferred to ferredoxin by NAD:ferredoxin oxidoreductase. If O_2 becomes available it serves as an electron acceptor and is reduced to H_2O_2 or to superoxide anion O_2^-.

One unusual feature of the function of hydrogenosomes is manifested by the hydrogenase (H_2:ferredoxin oxidoreductase). The enzyme provides a way for reoxidation of ferredoxin by the production of H_2. Hydrogenases have been identified in many bacteria as selenium- or nickel-containing proteins. Among eukaryotes *T. vaginalis* and *T. foetus* have hydrogenases that are located in the hydrogenosomes and play a vital function in reoxidation of ferredoxin. Electron paramagnetic resonance (EPR) spectroscopy studies have identified a [Fe]-containing hydrogenase in *T. vaginalis* that is highly sensitive to inhibition by carbon monoxide (Payne, Chapman, & Cammack, 1993).

More information is now available about the nature of the ferredoxin present in the hydrogenosomes. In addition to its function as an electron carrier for oxidative decarboxylation of pyruvate catalyzed by PFOR it also acts as the electron donor for the hydrogenase reaction. The hydrogenosomal ferredoxin from *T. vaginalis* was cloned and analyzed (Johnson *et al.*, 1990). The protein is composed of 93 amino acids and its primary structure confirms a close relationship to the [2Fe–2S] cluster and ferredoxins from several photosynthetic organisms and halobacteria. The gene is devoid of introns and is present in a single copy. The gene has an additional 8 amino acids at the amino terminus that are thought to be a leader sequence that is removed upon translocation into the hydrogenosome. Presumably, this is the site for attachment to the hydrogenosome inner membrane.

Pyruvate:Ferredoxin Oxidoreductase

The biochemical function of PFOR in the amitochondrial protists is analogous to that of the mammalian mitochondrial pyruvic dehydrogenase complex. The mammalian enzyme consists of four reactions catalyzed by three enzymes. The product of the oxidative decarboxylation of pyruvate in mammals ends with the formation of acetyl CoA, CO_2, and NADH. Subsequently, a great deal of energy is produced aerobically as a result of acetylCoA entering the citric acid cycle. The only catalytic mechanism that is shared by the mammalian pyruvic dehydrogenase complex and parasite PFOR is the thiamine pyrophosphate-dependent decarboxylation of pyruvate. The absence of lipoic acid, an active participant in the mammalian pyruvate dehydrogenase, in PFOR enzyme systems in the parasites, indicates a fundamental difference between parasite PFOR and its mammalian counterpart. PFOR plays a critical role in energy metabolism for all amitochondrial parasites, but it has no counterpart in mammals. PFOR should be considered an important target for prospective chemotherapeutic agents that interfere with vital parasitic energy metabolism without affecting the host.

Although recent studies on the nature of PFORs from representative parasites showed some similarities there are also significant differences. The enzyme from *E. histolytica* has been cloned and its primary structure has been identified (Rodriguez *et al.*, 1996). The amebic gene is 45% identical to the corresponding gene from *T. vaginalis* and 35% identical to that from *G. intestinalis*. Southern blot analysis showed that the *E. histolytica* gene exists only as a single copy. *Giardia*, however, has two different types of PFOR (Townson, Upcroft, & Upcroft, 1996). The first uses pyruvate as its preferred substrate and donates electrons to only one of three ferredoxins discovered (Fd I). The second oxidoreductase utilizes alternative substrates in addition to pyruvate with its preference being α-ketobutyrate. It is still not known whether this second oxidoreductase can donate electrons to any of the parasite ferredoxins. The availability of two different oxidoreductases raises the question whether there is any alternative electron transport pathway in *Giardia*. Since one of the other oxidoreductases has a preference for α-ketobutyrate, does that mean that *Giardia* has the metabolic flexibility to use an alternative to the pyruvate pathway? The answer is important, particularly in relation to metronidazole resistance.

Mechanism of Antiprotozoal Effects of Metronidazole and Related 5-Nitroimidazoles

The group of drugs that are effective against the three main amitochondrial protists are the 5-nitroimidazoles. These include metronidazole and tinidazole.

METRONIDAZOLE   TINIDAZOLE

Metronidazole (1-(2-hydroxyethyl)-2-methyl-5-nitroimidazole) was the first to be used for the treatment of *Trichomonas vaginalis* (Cosar & Julou, 1959) and is still the most commonly used. Since then, 5-nitroimidazoles were found to have a broad antimicrobial spectrum. In addition to vaginal trichomoniasis, metronidazole is the drug of choice against intestinal amebiasis and intestinal giardiasis. The 5-nitroimidazoles are also used against infections caused by anaerobic bacteria including *Bacteroides*, *Clostridium*, *Eubacteria*, and *Helicobacter*. Since the introduction of metronidazole, other new 5-nitroimidazoles such as tinidazole and ornidazole have been developed.

The antimicrobial effect of metronidazole and its congeners depends on the presence of a 5-nitro group on the imidazole ring. Derivatives without a nitro group do not have antimicrobial effects. Metronidazole is actually a "pro-drug" that has to be reduced by the microbes' biochemical machinery into the active antimicrobial form (Lindmark & Muller, 1976). Metronidazole, before reduction by the parasites, has no toxic effect against the microbes discussed here. The source of electrons for the reduction of metronidazole is the oxidative decarboxylation of pyruvate to acetyl CoA by pyruvate:ferredoxin oxidoreductase. The electrons are transferred to ferredoxin, because of its very low redox potential. Key reactions in the metabolism of pyruvate and their relationship to reduction of metronidazole are shown in Figs. 6.1 and 6.2. Once the nitroimidazole enters the parasite cell the nitro group, which has a low redox potential, accepts the electrons from reduced ferredoxin (Muller, 1986). The nitro group of metronidazole may accept up to four electrons to be converted to a stable amino group of the corresponding hydroxylamine. Neither the unreduced original 5-nitroimidazole molecule nor the final product of its reduction is toxic to the parasites. The cytotoxic effect appears to be due to the reductive activation of metronidazole to form labile chemically reactive anion radical derivatives (Moreno *et al.*, 1983; Muller, 1986; Docampo, 1990). Many attempts have been made to isolate and characterize the intermediary derivatives of reduced 5-nitroimidazole. Their instability has proven to be a great obstacle in studying their toxicity to the parasite. As shown in Fig. 6.3 free-radical intermediates may include the formation of a nitro-free anion radical of metronidazole (b) which was postulated to dismutate to a nitroso derivative (c) and then to a nitroso-free radical derivative (d), which is ultimately converted to a hydroxylamine (e) (Goldman *et al.*, 1986; Johnson *et al.*, 1990; Johnson, 1993). Oxygen can reverse the free radical to the original form of the drug (f). Studies using electron spin resonance spectroscopy (Lloyd & Pedersen, 1985) measured the steady-state intracellular levels of radical anions of metronidazole generated metabolically in intact *T. vaginalis*. These studies were also done in isolated enzyme systems containing pyruvate, CoA, and PFOR. Under anaerobic

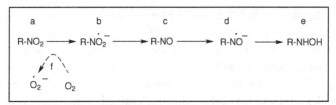

Figure 6.3. Reduction of the 5-nitro moiety of metronidazole (a) by the electrons transferred from ferredoxin to form nitro-free radical (b), nitroso (c), nitroso-free radical (d), and ultimately hydroxylamine (e). Under aerobic conditions, reoxidation (f) of the nitro-free radical produces the inactive form of the drug and a superoxide anion. Prepared from Johnson (1993) with permission.

conditions with N_2 equilibration, addition of metronidazole initiates maximal radical signal intensity, representing nitroso derivative b and its disappearance (Fig. 6.3). Stepwise increases in O_2 partial pressures resulted in a gradual decrease of the free radical signal concomitant with a decrease in the cytotoxic effect of metronidazole. Restoration of anaerobic conditions partially restores the recovery of free-radical production (Moreno et al., 1983; Muller, 1986; Docampo, 1990). There is also a difference between metronidazole-sensitive and resistant strains of T. vaginalis in their response to an increase in O_2. Sensitive organisms showed greater persistence of the free radicals produced by metronidazole than did the resistant strains. The free radical signals disappeared faster and at a lower O_2 tension in the resistant parasites. This shows that metronidazole free radicals can be reoxidized to inactive metronidazole. For a review see Johnson (1993). It has been found, however, that in the presence of low oxygen tension in the gut (Giardia) and in genitalia (Trichomonas), there is still enough reduced ferredoxin in the parasites to reduce the nitroimidazole, although at reduced efficiency.

So far, these 5-nitroimidazoles were shown to undergo reductive activation by the parasite's metabolic system before they can act on protozoal targets. The next obvious question is: What is the prime target or targets that the reduced form of the drugs act upon in the parasites? Although the molecular targets responsible for the cytotoxic effects of the reduced radical derivatives of nitroimidazoles have not been characterized, DNA was identified as one of these targets. Metronidazole inhibits the synthesis of new DNA in the parasite and causes degradation of existing DNA. Presumably, the reduced free-radical derivatives alkylate DNA with cytotoxic effects to the parasites. Damage appears to cause helix destabilization and strand breakage of DNA (Edwards, 1986). Damage is maximal with poly(d[AT]) but poly(d[GC]) is unaffected. DNA damage appears to be produced by short-lived reduction intermediates of nitroimidazoles. This is supported by the finding that DNA damage occurs only when reduction is carried out in the presence of DNA and not when DNA is added a few seconds later (LaRusso *et al.*, 1977). The final amino derivative of the reduction process is not cytotoxic and does not damage DNA. The evidence, therefore, supports the conclusion that both damage to the DNA and cytotoxicity is caused by the short-lived reduced intermediates of the nitroimidazoles. Certain proteins and membranes are also thought to be damaged by the reduced nitroimidazoles, but these targets have not yet been well defined.

Resistance to Metronidazole

Failure to cure human cases of giardiasis or trichomoniasis with metronidazole has been attributed to drug resistance. The minimum lethal concentration measured under aerobic conditions for metronidazole-sensitive isolates of *T. vaginalis* was 24.1 μg/ml whereas in isolates from treatment-resistant parasites it was 195.5 μg/ml. Corresponding mean anaerobic values were 1.6 and 5.05 μg/ml respectively (Muller, Lossick, & Gorrell, 1988). Clinical isolates of *T. vaginalis* that exhibit lower sensitivity to metronidazole were shown to have diminished capacity to reduce metronidazole (Lloyd, Yarlett, & Yarlett, 1986). Increased levels of resistance in these parasites correlate with decreased activity of pyruvate:ferredoxin oxidoreductase and ferredoxin (Johnson, 1993). An explanation for a decrease in the level of reduced ferredoxin was proposed by Quon, d'Oliveira, & Johnson (1992). Their data comparing sensitive with resistant strains are consistent with altered regulation of ferredoxin gene transcription. Using polyclonal antiserum raised against *T. vaginalis* ferredoxin they showed that the intracellular levels of this protein were significantly reduced in all resistant strains when compared with sensitive strains. Furthermore, levels of ferredoxin mRNA in resistant strains were reduced. This effect appears to be selective for ferredoxin mRNA since mRNAs expressing other proteins were not affected by changes in the sensitivity to metronidazole. The lower levels of ferredoxin mRNA are the result of a decrease in ferredoxin gene transcription. Comparison of sequence analysis of the 5' region of the ferredoxin gene uncovered two point mutations in the resistant strain. A protein of ~23 kDa was found to bind to the DNA region of one of these mutation sites. The binding affinity of this protein was found to be reduced in the mutants. It appears that drug resistance is related to altered regulation of ferredoxin gene transcription, resulting in less ferredoxin, which, in turn, reduces the ability of the protozoa to reduce metronidazole to its reactive intermediates (Quon *et al.*, 1992).

There is also experimental evidence that resistant strains of *T. vaginalis* have an alternative energy pathway that the organisms can use to avoid reduction of metronidazole. Brown *et al.* (1999) recently reported that although PFOR is down-regulated in resistant strains, there are two different 2-keto acid oxidoreductases that are fully active. These enzymes can metabolize pyruvate as well as α-ketobutyrate and α-ketomalonate as substrates, but they do not donate the electrons to ferredoxin nor to metronidazole (Brown *et al.*, 1999).

Clinical isolates of *Giardia intestinalis* showed considerable variation in their sensitivity to metronidazole ranging from 0.01 to 1 μg/ml (Upcroft & Upcroft, 1993). In the laboratory, by a process of stepwise increase in the concentration of metronidazole and selection of surviving parasites, *Giardia* can be made to tolerate 400 times the concentration of metronidazole lethal to sensitive strains. Among resistant strains of *G. intestinalis* there is a correlation between resistance to metronidazole and reduced levels of PFOR and to reduced production of toxic free radicals (Upcroft & Upcroft, 1993). Thus, the mechanism of resistance of *Giardia* to metronidazole appears to be similar to that for *Trichomonas*: a reduction in the ability of the parasite to reduce the antiprotozoal agent (Liu *et al.*, 2000).

Entamoeba histolytica is being discussed at the end of this section, although, clinically, it is the most important amitochondrial protist. The reason is that there have been no instances of clinical resistance to metronidazole reported. Furthermore, it has not been possible to develop resistant strains of amebae in the laboratory until recently. The induction of metronidazole resistance in axenically cultured trophozoites of *E. histolytica* following continuous exposure over 177 days to concentrations of metronidazole increasing from 1 to 10 μM has been reported (Samarawickrema *et al.*, 1997). More recently, the same group was able to isolate resistant strains that multiply in the presence of 40-μM metronidazole (Wassmann *et al.*, 1999). In contrast to mechanisms of metronidazole resistance in *Giardia* and *Trichomonas*, resistant amebae show no substantial downregulation of PFOR. Instead, there is a marked increase in the iron-containing superoxide dismutase (Fe-SOD). This is the enzyme that catalyzes the dismutation of superoxide radical anions to oxygen and hydrogen peroxide. Fe-SOD is known to act in conditions of generalized stress response. It also could be used by the resistant amebae for the detoxification of nitro free radicals derived from metronidazole activation in the parasites (Samarawickrema *et al.*, 1997). Besides the increase in Fe-SOD expression in resistant strains of amebae there is also increased expression of peroxiredoxin. Peroxiredoxin belongs to a recently discovered family of antioxidant enzymes that act as peroxidases, reducing H_2O_2 and alkyl hydroperoxides to water or the corresponding alcohol respectively. For a review see Schroder & Ponting (1998). Amebae do not contain detectable amounts of catalase or peroxidase that can scavenge hydrogen peroxide and other alkyl peroxides. These results were confirmed by episomal transfection of the antioxidant enzymes (Wassmann *et al.*, 1999). Reduction in the sensitivity to metronidazole of amebae was highest in cells simultaneously overexpressing Fe-SOD and peroxiredoxin. In summary, a mechanism has been described that is different from that in *Trichomonas* and *Giardia*. Sensitivity to metronidazole is lost by a combination of two enzymes, Fe-SOD and peroxiredoxin. The Fe-SOD and peroxiredoxin are presumably involved in detoxification of superoxide radical anions of reactive metronidazole metabolites.

Potentials for New Drug Targets

In the previous section the mechanism of action of metronidazole and its congeners on three representative amitochondrial protozoa, *Entamoeba*, *Giardia*, and *Trichomonas*, was described. Without doubt, metronidazole is the drug of choice against all these parasites because of its therapeutic effectiveness and its low toxicity. However, a good case can be made for searching for new chemotherapeutic agents against these organisms instead of depending only on metronidazole. For modern medical practice to depend on one drug for the treatment of amebiasis, giardiasis, and trichomoniasis is not foresighted. Although metronidazole is very effective against tissue amebiasis it is less effective against luminal amebiasis (parasites in the intestinal lumen). Emetine and dehydroemetine are drugs that were replaced by metronidazole 25 years ago. Although both of these drugs are still available, they are not recommended because of their possible serious cardiovascular toxicity. Patients using them require medical supervision, including electrocardiographic monitoring. Drugs such as diloxanide furoate and iodoquinol derivatives, which act on luminal amebiasis, are not therapeutically very effective and are reported to cause neuropathic toxicity and gastrointestinal complications, respectively. Additionally, in *Giardia* (Upcroft & Upcroft, 1993) and *Trichomonas* (Kulda, 1999; Sobel, Nagappan, & Nyirjesy, 1999) there is an increasing incidence of strains resistant to metronidazole.

With the recent advances in our knowledge about the unique metabolism of the amitochondrial parasites and the identification of many of the rate-limiting reactions essential for their survival, the time has come to identify new targets and new inhibitors that could increase our therapeutic armamentarium against these parasites.

Pyrophosphate Kinases

Phosphofructokinase (PP$_i$-PFK). One of the most distinctive features of glucose metabolism in the amitochondrial protists is their use of PP$_i$ in their energy metabolic pathways. Several kinases have been identified in *E. histolytica*, *G. intestinalis*, and *T. vaginalis* that use pyrophosphate as a phosphoryl donor (see previous section). These enzymes were shown to have different regulatory properties when compared to the corresponding ATP-dependent enzymes.

Pyrophosphate Analogs. Eubank and Reeves were the first to test PP_i analogs against ameba PP_i-PFK (Eubank & Reeves, 1982). One of these analogs, 1-hydroxynonane 1,1-diphosphonate inhibited PP_i-PFK activity and suppressed parasite multiplication *in vitro* at 1 mM concentration. The kinetics of inhibition were proven to be competitive with respect to PP_i, which indicates that the binding of this inhibitor involves the active site. Although a 1 mM concentration is considered to be rather high, this was an incentive to chemically modify the inhibitor so that affinity to the parasite enzyme would be increased.

In general, phosphonates that have a P–C–P structure rather than the P–O–P were found to be especially effective as inhibitors of the purified recombinant PP_i-PFK. Having a carbon atom replace the oxygen atom found in PP_i gives these compounds a more robust structure. All bisphosphonates tested were competitive inhibitors of PP_i-PFK but did not inhibit the mammalian ATP-PFK. Experiments on the effect of some of these inhibitors on amebic growth showed that 2-(3-pyridinyl)-1-hydroxyethylidene-1,1-bisphosphonate (risedronate) was the most effective inhibitor at a concentration of 10 μM. The low concentration at which these bisphosphonates inhibit amebic growth indicates that they might have a future as chemotherapeutic agents against amebiasis. They can probably best be used orally against intestinal lumen infections because of their poor absorption from the intestine (Bruchhaus *et al.*, 1996).

Several phosphonic acid derivatives, analogs of pyrophosphate, were examined against purified PP_i-PFK from the mitochondrial protist *Toxoplasma gondii* as well as against replication of *T. gondii* cultured in human foreskin fibroblasts (Peng *et al.*, 1995). Tetraisopropyl carbonyldiphosphonate 2,4 dinitrophenylhydrazone inhibited PP_i-PFK and also showed some selective inhibitory effect against replication of the parasites and helped protect the fibroblasts from

TETRAISOPROPYL CARBONYLDIPHOSPHONATE
2,4-DINITROPHENYLHYDRAZONE

damage. The structure of this compound is amenable to chemical manipulation and may have better access to protists that live in the lumen of the intestine (such as *Giardia*) than to intracellular parasites like *T. gondii*.

The possibility of using PP$_i$-PFK as a target for prospective amebicidal agents has been enhanced by cloning the gene from *E. histolytica* and expressing it in *E. coli* (Deng *et al.*, 1998). The properties of the expressed gene were found to be identical with the native PP$_i$-PFK purified from axenically grown *E. histolytica*. Because of the very high specific activity of the PP$_i$-PFK prepared by the Kemp group it is now accepted that this is the enzyme that plays a major role in *E. histolytica*. PP$_i$-PFK has also been identified in other protists (for reference see Mertens (1993)). Even in the absence of crystallographic data for a drug target, computer-assisted homology modeling has been used to predict structures of the target. Agabian and her colleagues (Byington *et al.*, 1997) have been able to use computer-assisted modeling with a 3-D model from the homologous structure of ATP-PFK from *B. stearothermophilus* (Evans, Farrants, & Hudson, 1981) to assist in predicting the structures of related enzymes. A computer program (DOCK 3.5) (Meng *et al.*, 1993) is being used to investigate the binding of pyrophosphate and PP$_i$ analogs with likely inhibitory effect on the enzyme based on the cDNA of *E. histolytica*. This basic model could also be used to screen thousands of other chemical compounds.

Pyruvate Phosphate Dikinase (PPDK). Another PP$_i$-dependent enzyme that has been identified in *E. histolytica* is pyruvate phosphate dikinase (PPDK), an enzyme that replaces pyruvate kinase. The enzyme has been cloned and expressed in *E. coli* (Saavedra-Lira, Ramirez-Silva, & Perez-Montfort, 1998). The recombinant enzyme was purified and the inhibitory effect of different PP$_i$ analogs was examined. Tetrasodium 1-hydroxymethylidene bisphosphonate was shown to be the most potent inhibitor ($K_i = 1.3$ mM) of the recombinant parasite PPDK. The same inhibitor was the most potent blocker ($K_i = 4.3$ mM) of the capacity of the amebae to lyse mammalian cells. PPDK from *Giardia intestinalis* was also characterized, but the effect of inhibitors on the enzyme was not examined (Hrdy, Mertens, & Nohynkova, 1993). Although these initial results with bisphosphonate inhibitors of PPDK were not remarkable they serve as a good starting point for designing a rational strategy to identify new amebicides. Future availability of a 3-D structure of PPDK will help to identify more selective inhibitors of this enzyme (Herzberg *et al.*, 1996).

Multisubstrate Analogs as Inhibitors of Pyrophosphate-Dependent Kinases. In spite of recent progress in molecular biology that allowed isolation of drug targets in sufficient amounts for studies on their physical and molecular characteristics there has not been enough interest in introducing new concepts for the chemical design of the most effective inhibitors. It has generally been accepted that the affinity and specificity of a ligand to bind to a target is enhanced by increasing the substrate binding area. In the case of enzymes or receptors that can bind more than one substrate or ligand, an inhibitor that combines properties of each substrate would be expected to have greater specificity against the target. This type of inhibitor is referred to as a multisubstrate analog inhibitor (MAI) (Broom, 1989).

A classical example of this kind of inhibitor is N-phosphonacetyl-L-aspartate (PALA), a selective inhibitor of aspartate carbamoyl transferase (Collins & Stark, 1969, 1971; see also Chapter 1). PALA resembles the two-substrate complex and causes large changes in the quaternary structure of the enzyme, thus favoring a transition state that has affinity to the inhibitor at nanomolar levels (Krause, Volz, & Lipscomb, 1985; Volz, Krause, & Lipscomb, 1986).

Several PP_i-PFKs of amitochondrial protists have been cloned and their active sites have been identified. It should be possible now to study more fully the binding interactions between the enzyme's active site and its two substrates, fructose 6-P and PP_i. Potential inhibitors could be designed on the basis of incorporation of the fructose 6-P as well as the pyrophosphate moieties to form a bisubstrate analog. Information already available on the inhibitory effect of some carbonyl diphosphonic acid derivatives on PP_i enzymes from protists as well as other protozoal parasites such as *Toxoplasma gondii* (Peng et al., 1995) should be helpful in identifying new MAIs of these enzymes.

Alcohol Dehydrogenases and Identification of Antiparasitic Drugs by Complementation of *E. coli* Mutants

Screening of prospective antiamebic agents against enzyme targets in parasites is usually hampered by the high cost of *in vitro* cultures, in addition to the experimental difficulties of obtaining accurate quantification of the efficacy of a large number of chemicals. Complementation of *E. coli* mutants that are missing an enzyme that is a possible target in a parasite provides a good method for identification of new antiamebic agents. A potential target in the fermentation pathway of *E. histolytica* is alcohol dehydrogenase/acetaldehyde dehydrogenase (EhADH2) (Yang et al., 1994). (See Fig. 6.1.) The enzyme is a bifunctional NAD^+-linked enzyme that is responsible for catalyzing the two terminal reactions in the fermentation pathway of *E. histolytica* to produce ethanol (Lo & Reeves, 1978; Bruchhaus & Tannich, 1994). EhADH2 appears to be a good target since it is structurally different from any known eukaryotic alcohol dehydroge-nase or acetaldehyde dehydrogenase. EhADH2 is homologous to the *E. coli* aldehyde dehydrogenase enzyme and can complement the missing enzyme in a bacterial mutant. An EhADH2 gene was expressed in a mutant strain of *E. coli* that had its acetaldehyde dehydrogenase gene deleted. Expression of EhADH2 in *E. coli* restored the ability of the mutant strain to grow under anaerobic con-ditions (Yong *et al.*, 1996). This *E. coli* strain was used to screen compounds that inhibit the *E. histolytica* alcohol dehydrogenase. The screening for inhibitors to this enzyme on a large scale uses inhibition of bacterial growth anaerobically, but not aerobically, as a test for enzyme inhibition. An added advantage to the use of this method is that bacterial growth can be quantitated by the simple technique of measuring the optical density of liquid bacterial cultures. Com-pounds capable of inhibiting anaerobic growth, but not aerobic growth, of the complemented *E. coli* mutant are potential specific inhibitors of *E. histolytica* alcohol dehydrogenase. The procedure was successfully documented by finding that pyrazole, a known inhibitor of alcohol dehydrogenase, inhibited anaero-bic, but not aerobic, growth of the *E. coli* mutant that was complemented with *E. histolytica* alcohol dehydrogenase. In addition, inhibition of the enzyme target was verified by demonstrating that pyrazole inhibited *E. histolytica* trophozoite growth *in vitro*. Pyrazole would not be a good candidate for an antiamebic agent because of the high concentration needed to inhibit amebic growth and hence its toxicity to the host. However, the procedure is valid for large-scale screening of other inhibitors of amebic alcohol dehydrogenase that would be more potent and less toxic. The procedure is very simple and less expensive than growing *E. histolytica* in culture and measuring growth inhibition by counting viable trophozoites. It deserves further consideration in the search for new chemother-apeutic agents against these parasites.

PFOR as a Target

Among the metabolic reactions used by the amitochondrial protists for energy production, PFOR plays a critical role in the survival of the parasite. Mammals do not have PFOR. Its counterpart, pyruvate dehydrogenase complex, is quite different from PFOR. Differences between the two enzyme systems makes PFOR an attractive target. Although PFOR was shown to be essential for reduction of metronidazole, producing intermediary free nitro radicals that are toxic to the parasites, its status as the drug target is not yet conclusively proven.

Studies on the relationship between PFOR activity and metronidazole resis-tance are inconsistent and make it difficult to accept PFOR as a reliable target for attacking these parasites. PFOR activity in metronidazole-resistant strains of *Giardia* (Townson *et al.*, 1996) or *Trichomonas* (Johnson, 1993) are markedly reduced, but the parasites are able to survive in the presence of high concentrations of metronidazole. Wang's group succeeded in reducing PFOR expression in *G. intestinalis* by 46–80%, using a *Giardia* virus-mediated hammerhead ribozyme, but the parasite grew equally well under anaerobic or aerobic conditions (Dan, Wang, & Wang, 2000). Studies on metronidazole-resistant strains of *T. vaginalis* (Brown *et al.*, 1999) and of *Giardia* (Townson *et al.*, 1996) indicate that these organisms have alternative electron transport systems as well as alternative sources of energy. Obviously, our knowledge of the role of PFOR in energy metabolism is as yet incomplete. However, it is still important to identify PFOR inhibitors for different parasites. The effects of these inhibitors could unravel some of the ambiguities regarding the regulatory function of PFOR in the energy metabolism of these parasites.

In spite of the reports about alternative electron carriers or metabolic pathways attempts have been made to find selective inhibitors of PFOR. Experimental evidence indicates differences between the active sites of PFOR from *T. vaginalis* and its counterpart, pyruvic dehydrogenase complex from *E. coli* (Williams, Leadlay, & Lowe, 1990). Although both enzyme systems were inactivated by preincubation with pyruvate, only PFOR from *T. vaginalis* was irreversibly inhibited by hydroxypyruvate, whereas only the bacterial pyruvic dehydrogenase was inhibited by bromopyruvate. Thiamine thiothiazolone pyrophosphate, an analog of thiamine pyrophosphate (the prosthetic group of both enzymes), did not inhibit *T. vaginalis* PFOR, but it caused complete inhibition of *E. coli* pyruvate dehydrogenase. This difference has been attributed to strong binding of the thiamine pyrophosphate to PFOR. PFORs from the three model parasites – *E. histolytica* (Rodriguez *et al.*, 1996), *T. vaginalis* (Hrdy & Muller, 1995), and *G. intestinalis* (Dan *et al.*, 2000) – have been cloned and their structures have been identified. The availability of recombinant PFOR enzymes in adequate amounts will facilitate more studies on their nature and their 3-D structures. Better understanding of the ways that these parasites use to circumvent different inhibitors is needed.

Proteinases as Targets

As discussed in Chapters 4 (malaria) and 5 (trypanosomes) proteinases play a prime role in sustaining parasitic life. In the amitochondrial protists the majority of these enzymes are cysteine proteinases that are involved in degrading cells and tissues of the host, phagocytosis of the red blood cells, and feeding on bacteria from the colon. The virulence and pathogenesis of the parasites is dependent on the participation of these enzymes. In addition, the proteinases also are used in the parasites' evasion of the host immune system.

Among the amitochondrial protists the proteinases of *E. histolytica* have been most thoroughly studied. Two comprehensive reviews on the subject have appeared (Que & Reed, 2000; Rosenthal, 1999). The most significant work has been on cysteine proteinases in *E. histolytica*. Six *E. histolytica* genes have been identified and molecular analysis of the genes confirm their relationship to the papain proteinase family. The enzymes appear to be involved as virulence factors mediating tissue invasion. The ability of the parasite to carry out lysis of host cells is associated with the cysteine proteinases. These proteolytic enzymes are located either on the amebae cell surfaces or in subcellular vesicles. Two cysteine proteinase genes, ACP_1 and ACP_2, have been identified from pathogenic isolates of *E. histolytica*. ACP_2 correlates with increased proteinase expression and activity. Gene function also correlates with the ability of the parasites to destroy fibroblasts in cell culture and with other *in vitro* assays of virulence (Reed *et al.*, 1993). The cysteine proteinases also degrade Immunoglobulin A (IgA), which plays a role as a mediator of immune responses against intestinal pathogens (Kelsall & Ravdin, 1993). The presence of extracellular cysteine proteinases in amebic liver abscesses was shown by Stanley *et al.* (1995) who also found that the cysteine proteinase inhibitor (L-trans-epoxysuccinyl-L-leucylamido-(4-guanadino)-butane, abbreviated E-64) reduced liver abscess formation in SCID mice infected with *E. histolytica*. The role played by cysteine proteinases in virulence of the parasites and in protecting *E. histolytica* from antibodies indicates that these enzymes would make good targets for antiamebic agents.

Garlic (*Allium sativum*) has been used since ancient times against different kinds of intestinal infections including dysentery (Chevalier, 1996). Allicin [S-(2-propenyl)2-propene-1-sulfinothioate], one of the active components of garlic, was reported to inhibit the ability of *E. histolytica* trophozoites in culture to destroy monolayers of mammalian cells (Mirelman, Monheit, & Varon, 1987). Recently, it was reported that allicin inhibits a cysteine proteinase in intact and lysed *E. histolytica* trophozoites at concentrations as low as 1 μM (Ankri *et al.*, 1997). The effect appears to involve sulfhydryl groups at the active site, because it was reversed with dithiothreitol. The results reintroduce an ancient chemical agent with antiprotozoal action that could be further exploited to identify more effective amebic proteinase inhibitors with higher chemotherapeutic titer.

The interest in proteinases has been extended to *Giardia* and *T. vaginalis*. *G. intestinalis* extracts show at least 18 proteolytic enzymes: These are predominantly cysteine proteinases but there are a few serine proteinases. Their M_r values are in the range of 30,000 to 211,000 (Williams & Coombs, 1995). A proteinase that is involved in excystation of *Giardia* was reported by the McKerrow group (Ward *et al.*, 1997). Excystation is a developmental step that includes rupture of the amebic cyst and the emergence of the replicating trophozoite into the small intestine, a key step in establishing infection in the host. This step is dependent upon a cysteine proteinase that is stored in peripheral vesicles of the trophozoites. Several inhibitors, including E-64 and some fluoromethyl ketones, blocked trophozoite excystation (Ward *et al.*, 1997). The *Giardia* cysteine proteinase gene was cloned, sequenced, and identified. This enzyme appears to be one of the earliest branches of the cathepsin B family and has evolved in primitive eukaryotic cells prior to the divergence of the plant and animal kingdoms (Ward *et al.*, 1997). From the evolutionary point of view these results indicate that parasites of the genus *Giardia* are one of the earliest lineages of eukaryotic cells.

T. vaginalis was also reported to have multiple cysteine proteinases. A number of these genes were identified (Mallinson *et al.*, 1994). The effects of cysteine inhibitors on trichomonad growth in axenic cultures were tested in parallel to the effects of the inhibitors on proteinase activity (Irvine, North, & Coombs, 1997). An inhibitor with broad specificity, E-64, had only a minor effect on the growth of the parasites but inhibited all but two (35 and 49 kDa) of the parasites' proteinases. Thus, not all of the proteinases are essential for growth in culture. However, a peptidyl acyloxymethyl ketone proteinase inhibitor (N-benzoyloxycarbonyl-Phe-Ala-Ch_2OCO-(2,6-$(CF_3)_2$)Ph) at micromolar levels killed *T. vaginalis*. At the same concentration it also inhibited the 35- and 49-kDa proteinases, the two enzymes that were not inhibited by E-64. These results indicate that one of the two proteinases is essential for growth of the trichomonads and could be a suitable target for new chemotherapeutic agents.

The study of proteinases of protozoal parasites is new, but much progress has been made. There remain many gaps in our basic knowledge of pathogenesis, proliferation, and immune system evasion. Although there has been, as yet, no new antiprotozoal agent discovered on the basis of antiproteinase effect to be used clinically, progress in this field gives hope that selective inhibitors of some of these enzymes would prove to be novel chemotherapeutic agents.

REFERENCES

Ankri, S., Miron, T., Rabinkov, A., Wilchek, M. & Mirelman, D. (1997). Allicin from garlic strongly inhibits cysteine proteinases and cytopathic effects of *Entamoeba histolytica*. *Antimicrob Agents Chemother, 41*(10), 2286–2288.

Broom, A. D. (1989). Rational design of enzyme inhibitors: Multisubstrate analogue inhibitors. *J Med Chem, 32*(1), 2–7.

Brown, D. M., Upcroft, J. A., Dodd, H. N., Chen, N. & Upcroft, P. (1999). Alternative 2-keto acid oxidoreductase activities in *Trichomonas vaginalis. Mol Biochem Parasitol, 98*(2), 203–214.

Bruchhaus, I. & Tannich, E. (1994). Purification and molecular characterization of the NAD(+)-dependent acetaldehyde/alcohol dehydrogenase from *Entamoeba histolytica*. *Biochem J, 303*(Pt 3), 743–748.

Bruchhaus, I., Jacobs, T., Denart, M. & Tannich, E. (1996). Pyrophosphate-dependent phosphofructokinase of *Entamoeba histolytica*: Molecular cloning, recombinant expression and inhibition by pyrophosphate analogues. *Biochem J, 316*(Pt 1), 57–63.

Byington, C. L., Dunbrack, R. L., Jr., Whitby, F. G., Cohen, F. E. & Agabian, N. (1997). *Entamoeba histolytica*: Computer-assisted modeling of phosphofructokinase for the prediction of broad-spectrum antiparasitic agents. *Exp Parasitol, 87*(3), 194–202.

Chevalier, A. (1996). *The Encyclopedia of Medicinal Plants* (pp. 21, 56). New York: DK Publishing, Inc.

Chi, A. S. & Kemp, R. G. (2000). The primordial high energy compound: ATP or inorganic pyrophosphate? *J Biol Chem, 275*(46), 35677–35679.

Collins, K. D. & Stark, G. R. (1969). Aspartate transcarbamylase. Studies of the catalytic subunit by ultraviolet difference spectroscopy. *J Biol Chem, 244*(7), 1869–1877.

Collins, K. D. & Stark, G. R. (1971). Aspartate transcarbamylase. Interaction with the transition state analogue N-(phosphonacetyl)-L-aspartate. *J Biol Chem, 246*(21), 6599–6605.

Coombs, G. & Muller, M. (1995). Energy metabolism in anaerobic protozoa. In J. Marr & M. Muller (Eds.), *Biochemistry and Molecular Biology of Parasites* (pp. 33–48). San Diego: Academic Press.

Cosar, C. & Julou, L. (1959). Activite de L'(hydroxy-2'ethyl)-1 methyl-2nitro-5 imidazole (8.823 R.P.) vis-a-vis des infections experimentales a *trichomonas vaginalis*. *Ann l'Institute Pasteur, 96*, 238–241.

Dan, M., Wang, A. L. & Wang, C. C. (2000). Inhibition of pyruvate-ferredoxin oxidoreductase gene expression in *Giardia lamblia* by a virus-mediated hammerhead ribozyme. *Mol Microbiol, 36*(2), 447–456.

Deng, Z., Huang, M., Singh, K., Albach, R. A., Latshaw, S. P., Chang, K. P. & Kemp, R. G. (1998). Cloning and expression of the gene for the active PPi-dependent phosphofructokinase of *Entamoeba histolytica*. *Biochem J, 329*(Pt 3), 659–664.

Diamond, L. S., Harlow, D. R. & Cunnick, C. C. (1978). A new medium for the axenic cultivation of *Entamoeba histolytica* and other *Entamoeba*. *Trans R Soc Trop Med Hyg, 72*(4), 431–432.

Docampo, R. (1990). Sensitivity of parasites to free radical damage by antiparasitic drugs. *Chem Biol Interact, 73*(1), 1–27.

Edwards, D. I. (1986). Reduction of nitroimidazoles in vitro and DNA damage. *Biochem Pharmacol, 35*(1), 53–58.

Edwards, M. R., Gilroy, F. V., Jimenez, B. M. & O'Sullivan, W. J. (1989). Alanine is a major end product of metabolism by *Giardia lamblia*: A proton nuclear magnetic resonance study. *Mol Biochem Parasitol, 37*(1), 19–26.

Edwards, M. R., Schofield, P. J., O'Sullivan, W. J. & Costello, M. (1992). Arginine metabolism during culture of *Giardia intestinalis*. *Mol Biochem Parasitol, 53*(1–2), 97–103.

Eubank, W. B. & Reeves, R. E. (1982). Analog inhibitors for the pyrophosphate-dependent phosphofructokinase of *Entamoeba histolytica* and their effect on culture growth. *J Parasitol, 68*(4), 599–602.

Evans, P. R., Farrants, G. W. & Hudson, P. J. (1981). Phosphofructokinase: Structure and control. *Phil Trans R Soc London B Biol Sci, 293*(1063), 53–62.

Finlay, B. & Fenchel, T. (1989). Hydrogenosomes in some anaerobic protozoa resemble mitochondria. *FEMS Microbiol Lett, 65,* 311–314.

Goldman, P., Koch, R. L., Yeung, T. C., Chrystal, E. J., Beaulieu, B. B., Jr., McLafferty, M. A. & Sudlow, G. (1986). Comparing the reduction of nitroimidazoles in bacteria and mammalian tissues and relating it to biological activity. *Biochem Pharmacol, 35*(1), 43–51.

Herzberg, O., Chen, C. C., Kapadia, G., McGuire, M., Carroll, L. J., Noh, S. J. & Dunaway-Mariano, D. (1996). Swiveling-domain mechanism for enzymatic phospho-transfer between remote reaction sites. *Proc Natl Acad Sci USA, 93*(7), 2652–2657.

Hrdy, I. & Muller, M. (1995). Primary structure and eubacterial relationships of the pyru-vate:ferredoxin oxidoreductase of the amitochondriate eukaryote *Trichomonas vaginalis. J Mol Evol, 41*(3), 388–396.

Hrdy, I., Mertens, E. & Nohynkova, E. (1993). Giardia intestinalis: Detection and char-acterization of a pyruvate phosphate dikinase. *Exp Parasitol, 76*(4), 438–441.

Hrdy, I., Mertens, E. & Van Schaftingen, E. (1993). Identification, purification and sepa-ration of different isozymes of NADP-specific malic enzyme from *Tritrichomonas foetus. Mol Biochem Parasitol, 57*(2), 253–260.

Irvine, J. W., North, M. J. & Coombs, G. H. (1997). Use of inhibitors to identify essential cysteine proteinases of *Trichomonas vaginalis. FEMS Microbiol Lett, 149*(1), 45–50.

Johnson, P. (1993). Metronidazole and drug resistance. *Parasitol Today, 9,* 183–186.

Johnson, P. J., d'Oliveira, C. E., Gorrell, T. E. & Muller, M. (1990). Molecular analysis of the hydrogenosomal ferredoxin of the anaerobic protist *Trichomonas vaginalis. Proc Natl Acad Sci USA, 87*(16), 6097–6101.

Kelsall, B. L. & Ravdin, J. I. (1993). Degradation of human IgA by *Entamoeba histolytica. J Infect Dis, 168*(5), 1319–1322.

Krause, K. L., Volz, K. W. & Lipscomb, W. N. (1985). Structure at 2.9-A resolu-tion of aspartate carbamoyltransferase complexed with the bisubstrate analogue N-(phosphonacetyl)-L-aspartate. *Proc Natl Acad Sci USA, 82*(6), 1643–1647.

Kulda, J. (1999). Trichomonads, hydrogenosomes and drug resistance. *Int J Parasitol, 29*(2), 199–212.

Kumar, A., Shen, P. S., Descoteaux, S., Pohl, J., Bailey, G. & Samuelson, J. (1992). Cloning and expression of an NADP(+)-dependent alcohol dehydrogenase gene of *Entamoeba histolytica. Proc Natl Acad Sci USA, 89*(21), 10188–10192.

LaRusso, N. F., Tomasz, M., Muller, M. & Lipman, R. (1977). Interaction of metronidazole with nucleic acids in vitro. *Mol Pharmacol, 13*(5), 872–882.

Lindmark, D. G. & Muller, M. (1976). Antitrichomonad action, mutagenicity, and reduc-tion of metronidazole and other nitroimidazoles. *Antimicrob Agents Chemother, 10*(3), 476–482.

Liu, S. M., Brown, D. M., O'Donoghue, P., Upcroft, P. & Upcroft, J. A. (2000). Ferredoxin involvement in metronidazole resistance of *Giardia duodenalis. Mol Biochem Parasitol, 108*(1), 137–140.

Lloyd, D. & Pedersen, J. Z. (1985). Metronidazole radical anion generation in vivo in *Trichomonas vaginalis*: Oxygen quenching is enhanced in a drug- resistant strain. *J Gen Microbiol, 131*(Pt 1), 87–92.

Lloyd, D., Yarlett, N. & Yarlett, N. C. (1986). Inhibition of hydrogen production in drug-resistant and susceptible *Trichomonas vaginalis* strains by a range of nitroimidazole derivatives. *Biochem Pharmacol*, *35*(1), 61–64.

Lo, H. S. & Reeves, R. E. (1978). Pyruvate-to-ethanol pathway in *Entamoeba histolytica*. *Biochem J*, *171*(1), 225–230.

Mallinson, D. J., Lockwood, B. C., Coombs, G. H. & North, M. J. (1994). Identification and molecular cloning of four cysteine proteinase genes from the pathogenic protozoon *Trichomonas vaginalis*. *Microbiology*, *140*(Pt 10), 2725–2735.

Martin, W. & Muller, M. (1998). The hydrogen hypothesis for the first eukaryote. *Nature*, *392*(6671), 37–41. [See comments.]

McLaughlin, J. & Aley, S. (1985). The biochemistry and functional morphology of the *Entamoeba*. *J Protozool*, *32*(2), 221–240.

McLaughlin, J., Lindmark, D. G. & Muller, M. (1978). Inorganic pyrophosphatase and nucleoside diphosphatase in the parasitic protozoon, *Entamoeba histolytica*. *Biochem Biophys Res Commun*, *82*(3), 913–920.

Meng, E. C., Gschwend, D. A., Blaney, J. M. & Kuntz, I. D. (1993). Orientational sampling and rigid-body minimization in molecular docking. *Proteins*, *17*(3), 266–278.

Mertens, E. (1993). Glycolysis revisited in parasitic protists. *Parasitol Today*, *9*, 122–126.

Mertens, E., Van Schaftingen, E. & Muller, M. (1992). Pyruvate kinase from *Trichomonas vaginalis*, an allosteric enzyme stimulated by ribose 5-phosphate and glycerate 3-phosphate. *Mol Biochem Parasitol*, *54*(1), 13–20.

Mertens, E., Ladror, U. S., Lee, J. A., Miretsky, A., Morris, A., Rozario, C., Kemp, R. G. & Muller, M. (1998). The pyrophosphate-dependent phosphofructokinase of the protist, *Trichomonas vaginalis*, and the evolutionary relationships of protist phosphofructokinases. *J Mol Evol*, *47*(6), 739–750.

Minotto, L., Tutticci, E. A., Bagnara, A. S., Schofield, P. J. & Edwards, M. R. (1999). Characterisation and expression of the carbamate kinase gene from *Giardia intestinalis*. *Mol Biochem Parasitol*, *98*(1), 43–51.

Mirelman, D., Monheit, D. & Varon, S. (1987). Inhibition of growth of *Entamoeba histolytica* by allicin, the active principle of garlic extract (*Allium sativum*). *J Infect Dis*, *156*(1), 243–244.

Moreno, S. N., Mason, R. P., Muniz, R. P., Cruz, F. S. & Docampo, R. (1983). Generation of free radicals from metronidazole and other nitroimidazoles by *Tritrichomonas* foetus. *J Biol Chem*, *258*(7), 4051–4054.

Muller, M. (1986). Reductive activation of nitroimidazoles in anaerobic microorganisms. *Biochem Pharmacol*, *35*(1), 37–41.

Muller, M. (1988). Energy metabolism of protozoa without mitochondria. *Annu Rev Microbiol*, *42*, 465–488.

Muller, M. (1993). The hydrogenosome. *J Gen Microbiology*, *139*, 2879–2889.

Muller, M., Lossick, J. G. & Gorrell, T. E. (1988). In vitro susceptibility of *Trichomonas vaginalis* to metronidazole and treatment outcome in vaginal trichomoniasis. *Sex Transm Dis*, *15*(1), 17–24.

Paget, T. A., Raynor, M. H., Shipp, D. W. & Lloyd, D. (1990). *Giardia lamblia* produces alanine anaerobically but not in the presence of oxygen. *Mol Biochem Parasitol*, *42*(1), 63–67.

Payne, M. J., Chapman, A. & Cammack, R. (1993). Evidence for an [Fe]-type hydrogenase in the parasitic protozoan *Trichomonas vaginalis*. *FEBS Lett*, *317*(1–2), 101–104.

Peng, Z. Y., Mansour, J. M., Araujo, F., Ju, J. Y., McKenna, C. E. & Mansour, T. E. (1995). Some phosphonic acid analogs as inhibitors of pyrophosphate-dependent phosphofructokinase, a novel target in *Toxoplasma gondii*. *Biochem Pharmacol, 49*(1), 105–113.

Que, X. & Reed, S. L. (2000). Cysteine proteinases and the pathogenesis of amebiasis. *Clin Microbiol Rev, 13*(2), 196–206.

Quon, D. V., d'Oliveira, C. E. & Johnson, P. J. (1992). Reduced transcription of the ferredoxin gene in metronidazole-resistant *Trichomonas vaginalis*. *Proc Natl Acad Sci USA, 89*(10), 4402–4406.

Reed, S., Bouvier, J., Pollack, A. S., Engel, J. C., Brown, M., Hirata, K., Que, X., Eakin, A., Hagblom, P., Gillin, F. *et al.* (1993). Cloning of a virulence factor of *Entamoeba histolytica*. Pathogenic strains possess a unique cysteine proteinase gene. *J Clin Invest, 91*(4), 1532–1540.

Reeves, R. E. (1984). Metabolism of *Entamoeba histolytica Schaudinn*, 1903. *Adv Parasitol, 23*, 105–142.

Reeves, R. E. & Guthrie, J. D. (1975). Acetate kinase (pyrophosphate). A fourth pyrophosphate-dependent kinase from *Entamoeba histolytica*. *Biochem Biophys Res Commun, 66*(4), 1389–1395.

Reeves, R. E., Menzies, R. A. & Hsu, D. S. (1968). The pyruvate-phosphate dikinase reaction. The fate of phosphate and the equilibrium. *J Biol Chem, 243*(20), 5486–5491.

Rodriguez, M., Hidalgo, M., Sanchez, T. & Orozco, E. (1996). Cloning and characterization of the *Entamoeba histolytica* pyruvate:ferredoxin osidoreductase gene. *Mol Biochem Parasitol, 78*, 273–277.

Rosenthal, P. J. (1999). Proteases of protozoan parasites. *Adv Parasitol, 43*, 105–159.

Saavedra-Lira, E., Ramirez-Silva, L. & Perez-Montfort, R. (1998). Expression and characterization of recombinant pyruvate phosphate dikinase from *Entamoeba histolytica*. *Biochim Biophys Acta, 1382*(1), 47–54.

Samarawickrema, N. A., Brown, D. M., Upcroft, J. A., Thammapalerd, N. & Upcroft, P. (1997). Involvement of superoxide dismutase and pyruvate:ferredoxin oxidoreductase in mechanisms of metronidazole resistance in *Entamoeba histolytica*. *J Antimicrob Chemother, 40*(6), 833–840.

Schofield, P. J., Edwards, M. R., Matthews, J. & Wilson, J. R. (1992). The pathway of arginine catabolism in *Giardia intestinalis*. *Mol Biochem Parasitol, 51*(1), 29–36.

Schroder, E. & Ponting, C. P. (1998). Evidence that peroxiredoxins are novel members of the thioredoxin fold superfamily. *Protein Sci, 7*(11), 2465–2468.

Sobel, J. D., Nagappan, V. & Nyirjesy, P. (1999). Metronidazole-resistant vaginal trichomoniasis – An emerging problem. *N Engl J Med, 341*(4), 292–293 [letter].

Stanley, S. L., Jr., Zhang, T., Rubin, D. & Li, E. (1995). Role of the *Entamoeba histolytica* cysteine proteinase in amebic liver abscess formation in severe combined immunodeficient mice. *Infect Immun, 63*(4), 1587–1590.

Strickland, G. T. (1991). Infections of the blood and reticuloendothelial system. In *Hunter's Tropical Medicine* (7th ed.). Philadelphia: Saunders.

Townson, S. M., Upcroft, J. A. & Upcroft, P. (1996). Characterisation and purification of pyruvate:ferredoxin oxidoreductase from *Giardia duodenalis*. *Mol Biochem Parasitol, 79*(2), 183–193.

Upcroft, J. & Upcroft, P. (1993). Drug resistance and *Giardia*. *Parasitol Today, 9*, 187–190.

Volz, K. W., Krause, K. L. & Lipscomb, W. N. (1986). The binding of N-(phosphonacetyl)-L-aspartate to aspartate carbamoyltransferase of *Escherichia coli*. *Biochem Biophys Res Commun, 136*(2), 822–826.

Ward, W., Alvarado, L., Rawlings, N. D., Engel, J. C., Franklin, C. & McKerrow, J. H. (1997). A primitive enzyme for a primitive cell: The protease required for excystation of *Giardia*. *Cell, 89*(3), 437–444.

Wassmann, C., Hellberg, A., Tannich, E. & Bruchhaus, I. (1999). Metronidazole resistance in the protozoan parasite *Entamoeba histolytica* is associated with increased expression of iron-containing superoxide dismutase and peroxiredoxin and decreased expression of ferredoxin 1 and flavin reductase. *J Biol Chem, 274*(37), 26051–26056.

Williams, A. G. & Coombs, G. H. (1995). Multiple protease activities in *Giardia intestinalis* trophozoites. *Int J Parasitol, 25*(7), 771–778.

Williams, K. P., Leadlay, P. F. & Lowe, P. N. (1990). Inhibition of pyruvate:ferredoxin oxidoreductase from *Trichomonas vaginalis* by pyruvate and its analogues. Comparison with the pyruvate decarboxylase component of the pyruvate dehydrogenase complex. *Biochem J, 268*(1), 69–75.

Yang, W., Li, E., Kairong, T. & Stanley, S. L., Jr. (1994). *Entamoeba histolytica* has an alcohol dehydrogenase homologous to the multifunctional adhE gene product of *Escherichia coli*. *Mol Biochem Parasitol, 64*(2), 253–260.

Yong, T. S., Li, E., Clark, D. & Stanley, S. L., Jr. (1996). Complementation of an *Escherichia coli* adhE mutant by the *Entamoeba histolytica* EhADH2 gene provides a method for the identification of new antiamebic drugs. *Proc Natl Acad Sci USA, 93*(13), 6464–6469.

7

Neuromuscular Structures and Microtubules as Targets

Parasites' ability to control their motility in their specific natural location in the host is critical to their survival. For example, *Ascaris* living in the small intestine of the host mammal have to move very vigorously to maintain their position and avoid expulsion with the flow of the intestinal contents. Any interference with coordination of the parasites' movements could result in their being carried to the large intestine, an environment that is hostile to *Ascaris* survival. Certain antiparasitic agents owe their effect to a selective inhibition of the motility of the parasites. These drugs can eliminate intestinal parasites as a consequence of their interference with the motility needed to maintain the parasites' position in the host.

Neuromuscular Physiology of Nematodes

Because of its large size and ready availability *Ascaris suum* is generally used as a prototype organism representing nematodes. It is generally accepted that there is a close relationship between the neuromuscular morphology and physiology of the free-living nematode *C. elegans* and *Ascaris*. For this reason *C. elegans* is frequently used as a prototype organism for experiments that involve genetic analysis or molecular biological manipulation. However, *Ascaris* has larger cells that are easier to access for electrophysiological experiments, such as electrode voltage clamp recordings. Pharmacological preparations made from portions of an intact *Ascaris* were used to measure muscle contractility by Baldwin as early as 1943 (Baldwin, 1943). Research on the structure and physiological function of the motoneuron system in *Ascaris* initiated by Stretton and his colleagues has shed more light on locomotive behavior and its control by motoneurons (Stretton *et al.*, 1978). For reviews see Johnson & Stretton (1980) and Davis & Stretton (1996). The nervous system of *Ascaris* has three interconnected parts.

A cephalic ganglia surrounds a pharyngeal nerve ring located at the anterior end of the body. From the pharyngeal nerve ring two main nerve cords, dorsal and ventral, descend; these are connected to smaller subdorsal and subventral cords. There are a number of motoneurons in the ventral cord that communicate with the dorsal nerve cord. Structure and physiological function of these motoneurons provide both inhibitory and excitatory effects on muscle contraction. There is another smaller set of ganglia in the posterior end of the parasite. The basic structure of the nervous system in *C. elegans* is similar to that in *Ascaris*.

The muscle cells of *Ascaris* have an unusual structure (Martin *et al.*, 1996). Instead of having a process of the nerve passing to the muscle there are processes of the muscle cells (called muscle arms) that pass to the nerves. This is achieved by a thin process that passes from the muscle cell to the syncytium, where it breaks into a number of finer processes known as "fingers." Each muscle cell has a contractile spindle region that lies under the hypodermis (the cell layer beneath the cuticle) and an enlarged balloon-shaped structure called the bag that contains the nucleus and glycogen particles, glycolytic enzymes, and mitochondria.

Nicotinic Acetylcholine Receptors (nAChRs)

Nicotinic acetylcholine receptors (nAChRs) are multimeric membrane-spanning proteins that belong to a large group of ligand-gated ion channels. These include receptors for γ-amino butyric acid (GABA), glycine (Gly), glutamate-gated chloride channels (only in invertebrates), and 5-hydroxytryptamine (5HT) (Karlin & Akabas, 1995). These channels play an important role in the regulation of the neuromuscular system. Binding of a specific ligand, such as acetylcholine or its agonists, results in opening of an ion channel that is selective for cations or anions.

The nicotinic acetylcholine receptor, in addition to recognizing acetylcholine as a neurotransmitter, is part of a channel in the membrane through which Na^+ and K^+ ions flow. The channel is formed of five subunits: two α subunits and one each of β, γ, and δ. For the channel to open, each of the two α subunits must bind one molecule of acetylcholine at the postsynaptic membrane. This increases the flow of Na^+ into the cell and the flow of K^+ out of the cell, which leads to depolarization of the postsynaptic membrane and initiates an action potential. The 3-D structure at 9-Å resolution of the nAChR from *Torpedo* has been determined (Unwin, 1993).

Bungarotoxin, the active principle from the venom of the krait, *Bungarus multicinctus*, antagonizes, almost irreversibly, neuromuscular transmission in the vertebrate skeletal muscle. It is used as a marker for identification of the nicotinic receptors in mammals, which are also identified pharmacologically by their responses to agonists (e.g., acetylcholine and carbamylcholine) or antagonists (e.g, d-tubocurarine and bungarotoxin).

Nicotinic Acetylcholine Receptors in Nematodes

Experimental evidence showing that *Ascaris* has acetylcholine receptors was first reported by Baldwin & Moyle (1949) using isolated muscle strip preparations. When acetylcholine is added to the saline fluid bathing the muscle strip it increases the tone of the spontaneous contractions and this is followed by a gradual cessation of contractility. Nicotine at concentrations lower than that of acetylcholine has a similar effect on *Ascaris* muscle strips. Subsequently, electrophysiological studies showed that acetylcholine causes depolarization as well as changes in spike frequency and amplitude (Del Castillo, De Mello, & Morales, 1963). Responses of *Ascaris* muscle to acetylcholine when measured either by contraction responses or changes in membrane potential are blocked by tubocurarine, but not by atropine. Therefore, the acetylcholine receptors in *Ascaris* are classified as nicotinic receptors. Attempts to further categorize these receptors on the basis of their responses to mammalian ganglionic nicotinic receptor antagonists and agonists were not successful. The nature of the *Ascaris* nicotinic receptors, therefore, appears to be unique to the nematodes and they have been designated nicotinic acetylcholine receptors of nematodes ($nAChR_n$; Martin *et al.*, 1996). The nematode receptors have specific agonists such as levamisole that act on the $nAChR_n$s, but they are less active against the homologous mammalian host receptors. The neuromuscular system of *Ascaris* contains the essential biochemical components of the cholinergic system. These include the presence of choline acetyl transferase (choline acetylase), the enzyme that synthesizes acetylcholine, and choline esterase, the enzyme that hydrolyzes the acetylcholine (Fleming *et al.*, 1996).

Identification of nAChRs from parasites has been facilitated by the success achieved in cloning homologous genes from *C. elegans*. Some of these genes are associated with anthelmintic action or resistance to anthelmintics. Progress with cloning of nAChR genes from parasites has been hampered by the difficulty in isolating genomic DNA and total RNA. Through the use of *C. elegans* genes several parasite genes have been cloned. These include nAChR subunits from the parasitic nematode *Onchocerca volvulus* (Ajuh & Egwang, 1994).

Comparison of nAChR subunits of nematodes with those of vertebrates shows some sequences that have been conserved in the nematodes. These include the portions of four transmembrane domains and the adjacent cysteine residues

(Cys 192 and Cys 193) in the α subunits that are critical for the binding of acetylcholine. However, there appear to be additional regions in the α subunits that have no counterparts in the vertebrate receptors. These are found closer to the N terminal than the adjacent cysteines (Fleming *et al.*, 1996). These differences between nematode nAChRs and those of vertebrates correspond with the pharmacological differences in their response to neurotoxins. For example, it has been reported that in *Ascaris* muscle and in *C. elegans* the κ-bungarotoxin is more effective as a nicotinic antagonist than the typical vertebrate muscle antagonist, α-bungarotoxin (Tornoe *et al.*, 1996). Another neurotoxin from soft coral that has been used to show the characteristic nature of the nematode nAChRs is lophotoxin and its analog, bipinnatin-B. It is a cyclic diterpene that has been shown to block all vertebrate and invertebrate muscle neuronal nAChRs tested (Abramson *et al.*, 1988, 1989). The toxins react covalently with tyrosine 190 of the α subunit of the *Torpedo* nicotinic receptor. This residue has been conserved in all α subunits tested that are sensitive to lophotoxin. The tyrosine is close to the adjacent cysteines that are critical for the acetylcholine binding site in the α subunits (Kao & Karlin, 1986). Tornoe *et al.* expressed the mouse muscle nicotinic receptors in *Xenopus laevis* oocytes and showed that bipinnatin B completely blocked mouse muscle nAChRs (Tornoe *et al.*, 1996). In contrast, *Ascaris suum* muscle cells containing the nAChR were found to be insensitive to high concentrations of bipinnatin B. Similarly, *C. elegans* nAChR expressed in *Xenopus* oocytes was not blocked by bipinnatin B. Although the tyrosine in *C. elegans* is in a slightly different position, this may or may not explain the lack of response to bipinnatin B inhibition. These results indicate that, in spite of the close homology between the sequences of α subunits from different species, they do vary in their responses to inhibitors. The pharmacological differences reported among the nAChRs, despite their partial homologies, give some indication that finding inhibitors to parasite nAChRs that do not inhibit host nAChRs is feasible.

Anthelmintics Acting Primarily on Nematode Nicotinic Receptors. The anthelmintics discussed here include the imidazothiazoles (levamisole), the tetrahydropyrimidines (pyrantel), the quaternary ammonium salts (bephenium), and the pyrimidines (methyridine). These compounds are referred to as nicotinic anthelmintics since their targets are the nicotinic receptors of the acetylcholine-gated membrane ion channels of the parasite. Levamisole was first introduced in veterinary practice in the 1960s as an anthelmintic against *Ascaris* and other gastrointestinal nematode infections. Subsequently, it was used briefly against nematode infections in humans. Pyrantel is effective against several human nematodes including *Ascaris*, the hookworms *Ancylostoma duodenale* and *Necator*

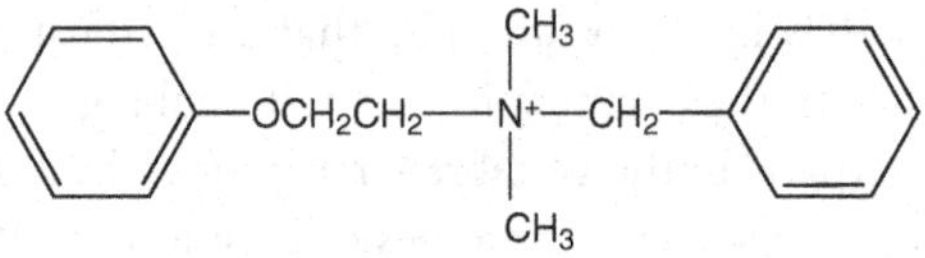

BEPHENIUM

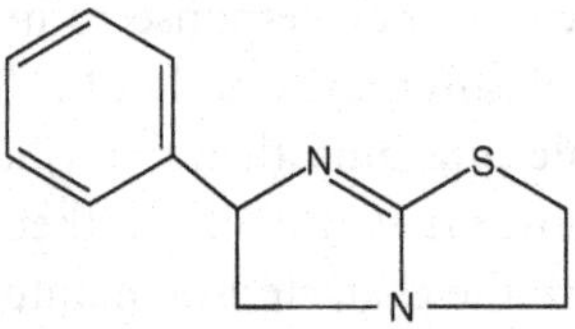

LEVAMISOLE (TETRAMISOLE)

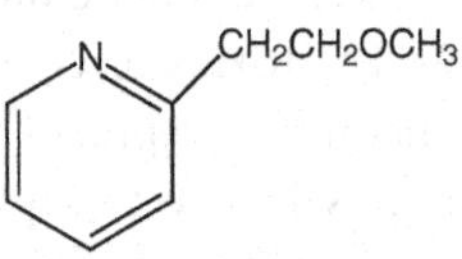

METHYRIDINE

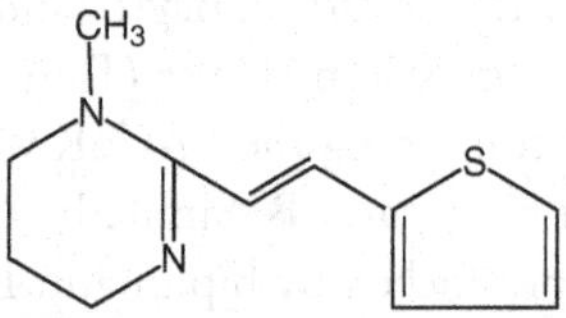

PYRANTEL

americanus, and the pinworm *Enterobius vermicularis*. The imidazothiazoles and other nicotinic anthelmintics act as agonists of the nicotinic acetylcholine receptors of *Ascaris* and other nematodes. They also have minor side effects on the hosts' nicotinic receptors. Pharmacological experiments on *Ascaris* muscle preparations demonstrated that levamisole induces neuromuscular inhibition and depolarization characterized by an increase in spike frequency. This leads to paralysis of the musculature of the parasite in the contracted state. Once paralyzed, the parasites are expelled with the gut contents from their natural habitat in the small intestine. These effects have been shown to occur with other nicotinic anthelmintics such as pyrantel, bephenium, and methyridine (Aceves, Erlij, & Martinez-Maranon, 1970; Van den Bossche, 1980).

Resistance to Levamisole. The development of resistance to levamisole and other nicotinic anthelmintics has become a major problem to animal breeders all over the world. A better understanding of the mechanism of resistance may contribute to identification of new drug targets and new diagnostic means to identify resistance. *Ascaris suum*, the model parasite for studying the mechanism of action of nicotinic anthelmintics, is not used for genetic studies on resistance. This is because *Ascaris* are usually obtained from the slaughterhouses and used mainly for short-term experiments. When studies on resistance that involve several generations of worms, are undertaken other nematodes, especially *C. elegans*, are preferred. *C. elegans* mutants that are highly resistant to levamisole and respond very poorly to related cholinergic agonists such as acetylcholine or nicotine have been isolated (Lewis *et al.*, 1980b). These results support previous conclusions based on pharmacological evidence that all members of the nicotinic anthelmintics, like levamisole, act on the nicotinic receptors. Genetic analysis showed that most of the resistant strains arise by mutation of any one of seven genes. Highly resistant mutants confer identical pharmacological phenotypes characterized by uncoordinated motor behavior and extreme resistance to cholinergic agonists (Lewis *et al.*, 1980a). Evidence that resistance in levamisole mutants is due to a change in the ability of their nicotinic receptors to bind levamisole was reported. Binding experiments using a potent levamisole derivative, tritium-labeled meta-amino levamisole, showed that extracts from wild-type worms have greater affinity for levamisole binding compared to resistant mutants (Lewis *et al.*, 1980a).

Although *C. elegans* and parasitic nematodes show close morphological and biological relationships, attempts are always made to confirm findings in *C. elegans* using a representative parasitic nematode. *Oesophagostomum dentatum* is a 1-cm-long parasitic nematode that lives in the large intestine of the pig. Resistant strains were selected for by treating pigs with increasing doses of levamisole and using the levamisole-resistant larvae to infect other pigs. Levamisole-resistant strains developed after 10 generations (Varady *et al.*, 1997). Martin and his associates used these parasites to study changes in the nicotinic receptors of levamisole-resistant and -sensitive isolates (Robertson, Bjorn, & Martin, 1999). Using a patch-clamp technique they found that the number of channels gated by 10 μM levamisole was similar in sensitive and resistant isolates. However, the number of channels gated by 30 μM and 100 μM levamisole were significantly reduced in resistant isolates. Also, the open times of the channels were reduced in resistant strains. The receptors of sensitive and resistant isolates were shown to be heterogeneous. Resistant isolates had a lower percentage of active receptors when the concentration of levamisole was 30 μM or higher. This confirms that a change in channel properties in the resistant isolates is responsible for resistance to levamisole. These may include changes in the agonist binding sites of levamisole or changes in the transmembrane regions of the receptor. It has been suggested that nicotinic receptors of *C. elegans*, and perhaps those of other nematodes, could comprise varying combinations of the different subunits (Martin, Robertson, & Bjorn, 1997). The nature of these receptors from resistant strains has not yet been determined.

Diethylcarbamazine, an Antifilarial Agent. Human filariasis is caused by nematode parasites that live in lymph glands, deep connective tissues, subcutaneous tissues, or mesenteries. *Wuchereria bancrofti* and *Brugia malayi* that are localized in the lymphatics are representative species. *Onchocerca volvulus* is located in the skin, eye, and lymphatics and could cause the clinical condition known as river blindness. Diethylcarbamazine, a piperazine derivative, is considered to be the most important chemotherapeutic agent against the microfilariae of *W. bancrofti* and *B. malayi* but not against the adult stages of the parasites. Diethylcarbamazine acts on the motility of the microfilariae, causing immobilization and subsequent elimination. Martin *et al.* carried out electrophysiological experiments on the effect of diethylcarbamazine on the membrane potential of *Ascaris suum* muscle cell preparations (Martin, 1982). Addition of acetylcholine to these preparations caused an increase in input conductance. This response was reversibly antagonized by d-tubocurarine and by diethylcarbamazine but not by piperazine. Diethylcarbamazine alone produced a reversible depolarization of the membrane potential with an increase in the amplitude and frequency of spontaneous depolarizing potentials. Further evidence for diethylcarbamazine's inhibitory effect on the cholinergic receptor was shown by its antagonistic effect on the voltage-sensitive outward current of the muscle bag preparation that is mimicked by a high K^+ level in the perfusion fluid. Immobilization of the microfilariae by diethylcarbamazine is thought to occur by a similar mechanism. After immobilization and dislocation from their normal habitat in the host the microfilariae appear to have a change in their surface membranes that renders them more susceptible to destruction by the host's immune system.

$$CON\,(C_2H_5)_2$$

DIETHYLCARBAMAZINE

γ-Amino Butyric Acid (GABA)

GABA is a neurotransmitter that is ubiquitous in vertebrate and invertebrate nervous systems. It is most abundant as an inhibitory transmitter. In vertebrates there are two main types of GABA receptors, GABA-A and GABA-B, and a GABA transporter. GABA-A is the most prevalent in mammalian central nervous systems and it is the best studied. The receptor is part of a transmembrane chloride ion channel. Like the acetylcholine-activated cation receptor the GABA receptor is made up of five transmembrane subunits that comprise a conduction channel across the membrane. Binding sites for GABA agonists or antagonists are located on the outer surface of the membrane. Binding of two GABA agonist molecules (such as muscimol) at the receptor increases chloride permeability. This decreases the depolarization of an excitatory impulse with a final outcome of depressing excitability. GABA-B receptors are identified at lower levels in the mammalian central nervous system and are not part of a chloride channel but are coupled to a G-protein. The GABA transporter is involved in the uptake of GABA in the brain.

GABA-A is the best-studied GABA receptor in parasites and is most related to the GABA-A mammalian receptor. Studies suggested that GABA is an inhibitory neuromuscular transmitter in *Ascaris* at neuron–muscle and neuron–neuron synapses (Del Castillo, De Mello, & Morales, 1964). Studies in *Ascaris* showed that their GABA receptors are pharmacologically different from vertebrate GABA-A receptors (Martin *et al.*, 1996). For example, the relative potency of a series of agonists on conductance–dose response relationships was found to be similar, but not identical, to that of the vertebrate GABA-A receptor. Furthermore, several vertebrate GABA-A competitive antagonists do not show the same rank order of potency when tested on *Ascaris* GABA receptors. These differences justify Martin's suggestion that there is a distinct type of GABA receptor in nematodes, identified as "GABA-N." The question of how GABA is taken up in *Ascaris* was studied by Guastella *et al.* Pharmacologically, the GABA uptake system for *Ascaris* is different from GABA uptake in mammals (Guastella & Stretton, 1991). For example, the uptake in *Ascaris* is sodium dependent but is not inhibited by the mammalian GABA uptake inhibitor nipecotic acid (a derivative of nicotinic acid). The distinctive nature of the nematode receptor is of importance because it offers a new target, in addition to the cholinergic receptor target in *Ascaris*, that leads to paralysis of the parasite. A combination of levamisole with a GABA receptor antagonist has been suggested by Martin and his associates to have a synergistic inhibitory effect on motility. The GABA-gated chloride channels that have been described so far are distinct from the glutamate-gated chloride channels that are the targets for ivermectin (see the following).

PIPERAZINE

The Action of Piperazine (Diethylenediamine) on GABA Receptors

Piperazine was the drug of choice against *Ascaris* and pinworms in children and adults until it was replaced by the benzimidazoles. Originally, it was thought that piperazine acts on *Ascaris* by binding to nicotinic receptors, causing paralysis of parasite musculature (Norton & DeBeer, 1957). More recently, Martin presented convincing data showing that piperazine owes its effect to binding to GABA receptors of *Ascaris* (Martin, 1982, 1985). Experiments to measure the opening and closing times and conductance state of the channel were done on the "bag" region of the somatic muscle of *Ascaris*, a region that is more accessible than other parts of the muscle cells for using a patch-clamp technique. Although piperazine has a heterocyclic ring structure that is quite different from GABA, it was shown that the two compounds act on the same receptors that are part of the GABA-gated Cl^- channels. Piperazine acts as a GABA agonist to open the chloride channels in the nematode somatic musculature. The outcome of this effect is an increase of Cl^- conductance at the somatic muscle membrane, leading to hyperpolarization and a reduction in contractility. The ultimate effect is flaccid paralysis of the muscle fibers. Consequently, the parasites are unable to maintain their position in the host and are eliminated with the gut contents. GABA is more than 10 to 100 times more potent than piperazine when conductance–dose responses are determined. However, GABA cannot be used as an anthelmintic instead of piperazine because it is highly ionized and could not cross the *Ascaris* cuticle to get to the muscle receptors.

Glutamate as a Neurotransmitter in Nematodes

In the mammalian central nervous system glutamate is recognized as an excitatory neurotransmitter that causes depolarization of the neurons. There are two classes of glutamate receptors that regulate channels permeable to Ca^{2+}, K^+, and

Na^+. Recent experimental evidence indicates that glutamic acid plays a similar role in neuronal functions in nematodes. Glutamate receptors have been identified in *C. elegans* and *Haemonchus contortus*. A glutamate receptor gene (*glr-1*) is expressed in motoneurons and interneurons of *C. elegans*, including those implicated in the control of locomotion (Maricq *et al.*, 1995). Furthermore, electrophysiological and pharmacological experiments showed that glutamate plays a role in nematode movement (Davis, 1998). When the worms are injected with glutamate or the agonist kainic acid, they become paralyzed. Kainate is the active principle from red algae, which has in the past been used in Asia as a potent anthelmintic (Takemoto, 1978). It is an important pharmacological test reagent for the study of glutamate receptors. The paralytic effect of glutamate or kainate is characterized by an apparent static anterior body posture. This effect is due to an action on glutamate receptors on dorsal excitor motoneurons. Studies on these receptors in parasitic nematodes have been limited.

In addition, nematodes have glutamate-gated chloride channels that are not found in vertebrates and are different from the glutamate channels described above (Cully *et al.*, 1996b). Similar channels were also found in insects, crustaceans, and mollusks. Neither examinations of recombinant DNA nor biochemical and physiological studies have discovered related gluatmate-gated chloride channels in vertebrates. Great interest in these channels arose when it was discovered that glutamate and ivermectin, an antinematodal agent, interact at the same chloride channels. Electrophysiological studies on *C. elegans* and *Ascaris suum* revealed that inhibitory motoneurons in these nematodes are controlled by the glutamate-gated chloride channels (Martin, 1996). The use of these glutamate receptors as targets in searching for new anthelmintics has yet to be explored.

Action of Avermectins on Glutamate-Gated Chloride Channels

The discovery of the avermectin group of anthelmintics began while testing fermentation products of a new species of actinomycetes, isolated from the soil by workers at Kitasata Institute in Japan (Campbell *et al.*, 1984). Soil samples and fermentation products were sent to Merck Institute of Therapeutic Research for screening on mice infected with *Nematospiroides dubius* (a parasitic *Saccharomycetaceae*). The actinomycetes species that had the best anthelmintic effect was identified and named *Streptomyces avermitilis* (the streptomyces that is capable of acting on worms). The anthelmintic activity was shown to reside in eight closely related macrocyclic lactones, named avermectins. Some of the avermectins were found to have activity against free-living and parasitic arthropods. A chemical

IVERMECTIN

derivative of one of these macrocyclic lactones, 22, 23-dihydroavermectin B1 or ivermectin, was developed because of its high efficacy as an antiparasitic agent. It has a broad spectrum of anthelmintic activity with low toxicity in mammals. Ivermectin is now widely used in veterinary medicine against a variety of intestinal nematodes. In humans it is used against the filaria *Onchocerca volvulus*, the causative organism of river blindness in Africa (Greene *et al.*, 1985). As an anthelmintic its use was restricted to a variety of gastrointestinal nematodes. In general, it is not effective against many trematode or cestode parasites. Abamectin, a combination of two avermectin derivatives, was designed as an insecticide and miticide and was shown to be effective against several species of ticks, mites, lice, and grubs (Campbell, 1985).

Ivermectin is a model antiparasitic agent that owes its effect to selective paralysis of parasite musculature. The effect is initiated through interaction of ivermectin with receptors on the glutamate-gated chloride channels. Ivermectin binding to the chloride channels results in an increase in chloride ion conductance, hyperpolarization, and paralysis of the musculature. Ivermectin also causes paralysis of the nematode pharynx and stops oral ingestion of nutrients. Because these chloride channels are present only in invertebrates the effect of ivermectin is selective against the nematodal parasites and thus they are considered to be ideal targets for antinematodal agents and insecticides.

Investigations into the molecular mechanism of action of avermectin analogs on nematodes were carried out using mRNA from the free-living nematode *C. elegans*. *Xenopus laevis* oocytes, when injected with the mRNA from *C. elegans*, expressed glutamate-gated chloride channels that were used to examine the mechanism of action of the avermectins. Using different avermectin analogs, a strong correlation was found between their nematodicidal potency and their binding to and activation of the glutamate-gated chloride channels (Arena *et al.*, 1995). The cDNAs for two glutamate-gated chloride channels were isolated from

C. elegans (designated *GluClα* and *GluClβ*; Cully *et al.*, 1994) and subsequently from the filarial nematodes *Dirofilaria immitis* and *Onchocerca volvulus* (Cully *et al.*, 1996b). Hydrophobicity analysis of these chloride channel proteins showed motifs common to all ligand-gated channels including those of *Drosophila melanogaster* (Cully *et al.*, 1996a). The pharmacological properties of the α and β glutamate-gated chloride channels from *C. elegans* were studied after their expression in *Xenopus* oocytes either together or individually. The evidence indicates that the two subunits GluClα and GluClβ are required to enable both glutamate and avermectin gating. When both α and β channels were expressed in the *Xenopus* oocytes as heteromers they showed rapid irreversible activation with ivermectin and rapid reversible activation with glutamate. However, when the α and the β chloride channels were expressed individually, forming homomeric channels, different results were obtained. Monomeric α channels expressed in *Xenopus* oocytes were sensitive to ivermectin but not glutamate, whereas oocytes expressing β channels were sensitive to glutamate but not ivermectin (Cully *et al.*, 1994). These results indicate that coexpression of GluClα and β in the oocytes leads to the formation of heteromeric channels. The channels in nematodes are heteromeric (containing both α and β subunits). Having both subunits led to ivermectin potentiation of the glutamate response (Cully *et al.*, 1996b).

Glutamate Receptors in the Pharynx of Nematodes

The effect of ivermectin in causing general worm paralysis appears to be insufficient to explain its observed lethal effect on nematodes. Pharyngeal muscles of nematodes are known to be involved in pumping food inside the esophagus and intestines of the parasite and are known to receive an inhibitory motoneuron. In *C. elegans* ivermectin inhibits pharyngeal pumping at drug concentrations too low to cause complete paralysis of other muscles (Avery & Horvitz, 1990). In the parasitic nematode *Haemonchus contortus* ivermectin inhibits ingestion of food at concentrations as low as 10^{-10} M (Geary *et al.*, 1993). This is lower than concentrations that impair motility (10^{-8} M) or carbohydrate metabolism (10^{-6} M). This experimental evidence indicates that the pharyngeal muscle is much more sensitive than the somatic muscles necessary for general parasite motility. In the hookworms, *Necator americanus* and *Ancylostoma ceylanicum*, ivermectin also inhibits pharyngeal pumping and food ingestion (Richards, Behnke, & Duce, 1995). These results drew attention to the nervous control of the pharynx of nematodes as a target for antiparasitic agents. Electrophysiological studies on isolated pharyngeal muscle preparations from *Ascaris suum* confirm that it possesses a glutamate-gated chloride channel that is sensitive to avermectin and its analog milbemycin D. The muscle required for pharyngeal pumping is known to receive a glutaminergic inhibitory motoneuron (M_3) (Avery, 1993). Application of glutamate to the pharyngeal muscle results in hyperpolarization and a dose-dependent increase in chloride ion conductance. The effect of glutamate is potentiated by the avermectin analog milbemycin D, indicating interaction with the same channel (Martin, 1996). In summary, the antiparasitic effect of the ivermectins appears to be the result of starvation and/or paralysis of the affected organisms.

Resistance to Ivermectin. Ivermectin and its congeners are used against nematode infections in domestic animals. Because of the widespread use of these anthelmintics, some parasites have developed resistance. Research on resistance to ivermectins focuses on *in vivo* and *in vitro* studies on several species of trichostrongylid nematodes. *Haemonchus contortus*, a stomach worm infection of sheep, is frequently used as a model parasite. An *in vitro* technique developed by Geary *et al.* (1993) uses fluorescent labeled *E. coli* to measure pharyngeal ingestion of *H. contortus*. Using this technique Sangster and his group developed ivermectin-resistant isolates that tolerated 177-fold higher concentrations of ivermectin than susceptible worms (Sangster *et al.*, 1999). In addition, motility of ivermectin-resistant parasites was shown to be less than that of susceptible parasites. The resistance of the worms does not appear to be caused by a change of binding properties of ivermectin to membrane preparations of *H. contortus*. Both affinity as well as receptor density appear to be similar when susceptible and resistant isolates are compared (Rohrer *et al.*, 1994). These results, although they do not indicate involvement of ivermectin receptors in the process of drug resistance, do not exclude a mechanism based on receptor involvement

Amplified expression of multidrug resistance genes is a mechanism used by cancer cells and *Plasmodia* to increase the efflux of different types of cytotoxic agents. These genes express transmembrane P-glycoproteins that promote the process of active drug efflux (see Chapter 4). The presence of P-glycoproteins in *C. elegans* was observed to provide protection against natural toxic compounds (Broeks *et al.*, 1995). The possibility that an increase in efflux of ivermectin can explain the mechanism of resistance in *H. contortus* has been investigated in two laboratories. The data from both laboratories suggest that a P-glycoprotein may play a role in ivermectin resistance. Expression of P-glycoprotein mRNA was found to be higher in resistant isolates compared to susceptible isolates of *H. contortus* (Xu *et al.*, 1998). Using the DNA of internucleotide binding domain of P-glycoproteins as a probe of Southern blots of *H. contortus* Sangster *et al.*

(1999) demonstrated a pattern consistent with the involvement of P-glycoprotein in resistance to ivermectin.

Bioamine Receptors in Nematodes

Several bioamine neurotransmitters have been identified in free-living and parasitic nematodes. These include serotonin, dopamine, epinephrine, nore-pinephrine, and octopamine. A review that deals with their physiological actions, syntheses, and metabolism has recently appeared (Brownlee & Fairweather, 1999). Because of its early discovery and its widespread distribution in many invertebrate species (Erspamer, 1954) and in parasites (Mansour, Lago, & Hawkins, 1957; Mansour, 1984) more attention was given to serotonin by investigators who were interested in the biology of parasitic helminths. In nematodes serotonin has been implicated in regulation of motility, pharyngeal contractions, mating and egg laying behavior, and as a hormone that regulates carbohydrate metabolism. Much of the serotonin research in nematodes has been done on *Ascaris suum* or *C. elegans*. This indolamine has been identified histochemically in *C. elegans* in whole organisms as well as in the pharyngeal neurons. Exogenous serotonin depresses motility but stimulates pharyngeal pumping in intact *C. elegans* (Horvitz et al., 1982). In isolated preparations of *Ascaris suum* the pharynx does not contract spontaneously, but serotonin added at micromolar concentrations activates rhythmical contractions and pharyngeal pumping (Brownlee, Holden-Dye, & Walker, 1997). In *C. elegans* control of motility by serotonin appears to be associated with heterotrimeric guanine nucleotide-binding proteins (G proteins). Mutants that have no serotonin receptors move hyperactively. Similarly, mutants that lack the gene that encodes for the α subunit of G_o have defects similar to those that lack serotonin receptors. Thus G_o activity is essential for the behavioral effect of serotonin on these nematodes (Segalat, Elkes, & Kaplan, 1995). The relationship between serotonin receptors and a second messenger should be further investigated. In addition to its involvement in the regulation of motility and behavior of nematodes, there is experimental data (see Chapter 3) to indicate that serotonin is linked to the control of glycogenolytic enzymes through cAMP in *Ascaris suum*.

Pharmacological and physiological data indicate that, as in vertebrates, a variety of receptors and neurons are involved in the control of motility and of carbohydrate metabolism of nematodes. A serotonin receptor from *C. elegans*, 5-HT-Ce, has been cloned and the gene was expressed in murine LtK-cells. This receptor shows the structural characteristics of seven putative transmembrane loops, which indicates that it may be a member of the family of receptors that is linked to a GTP-binding protein. The signal would then be transmitted to adenylate cyclase, which synthesizes the second messenger, cyclic-AMP. Serotonin inhibits adenylate cyclase and appears to function as a ligand that can constrain cyclic-AMP production by the cyclase. A clonal cell line (Ltk-cells) was constructed to permanently express the gene to facilitate research on the expressed receptor protein (Olde & McCombie, 1997). Furthermore, a 5-HT receptor was cloned from the pharynx of *Ascaris suum*. That receptor was expressed in COS-7 cells and was shown to bind lysergic acid diethylamide (LSD), an ergot alkaloid derivative and a serotonin agonist. As in the case of *C. elegans* the receptor has the structural characteristics of seven transmembrane loops, (Huang *et al.*, 1999). Serotonin receptors from both *C. elegans* and *Ascaris* exhibit high affinity binding to LSD and lower affinities for binding serotonin and related indolalkylamines. It is evident from the primary research carried out on *C. elegans* and *Ascaris* that the serotonin receptors are part of a signaling system that regulates carbohydrate metabolism and motility.

Microtubules as a Target for Benzimidazoles in Nematodes

The introduction of benzimidazoles in therapeutics as broad spectrum anthelmintics affecting a variety of different types of nematodes marked the beginning of a new era in the treatment of many gastrointestinal nematode infections (Brown *et al.*, 1961). The benzimidazoles owe their primary action to binding to the cytoskeletal protein, tubulin, and blocking the formation of the microtubular matrix. Microtubules are filaments arranged in a cylindrical shape made up of polymerized 50-kDa subunits of α- and β-tubulin. They extend across the cytoplasm of cells and control the location of membrane-bound organelles and other intracellular components. α- and β-tubulin are arranged in longitudinal rows to form a number of linear protofilaments. This is the arrangement that gives the microtubule a cylindrical shape with a hollow central core (Alberts *et al.*, 1994). One essential feature of the microtubules is their ability to polymerize and depolymerize, a process that is controlled by GTP/Mg^{2+}. There is always a dynamic equilibrium between free tubulin subunits and the "treadmilling" addition and subtraction of dimeric tubulin units occurring at opposite ends of the developing tubule (Lacey & Gill, 1994). There are two sites for GTP. One is in the β-tubulin (the + site) where the bound GTP is hydrolyzed to GDP during polymerization of tubulin. The rate of addition of tubulin is enhanced by GTP, Mg^{2+}, and MAPS (microtubular associated proteins) and decreased by Ca^{2+}, calmodulin, and different drugs. The other GTP site is found on the α-tubulin

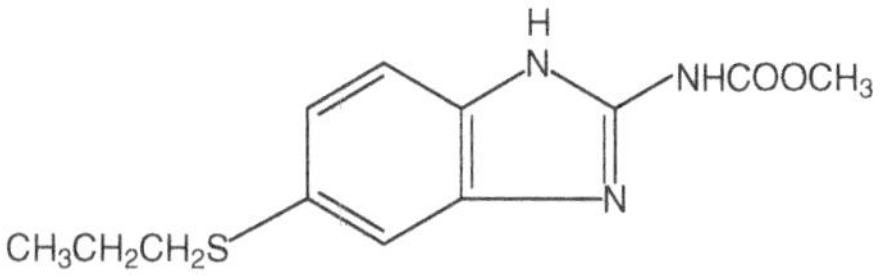

ALBENDAZOLE

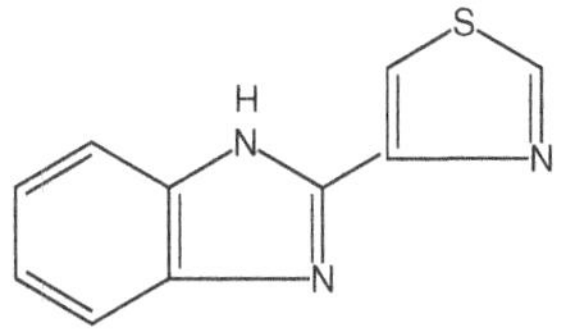

THIABENDAZOLE

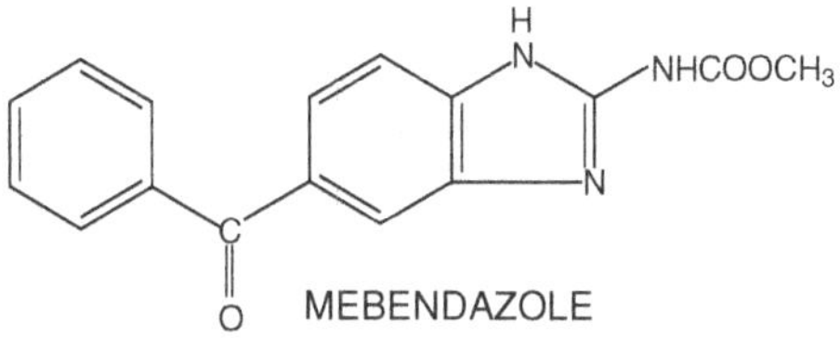

MEBENDAZOLE

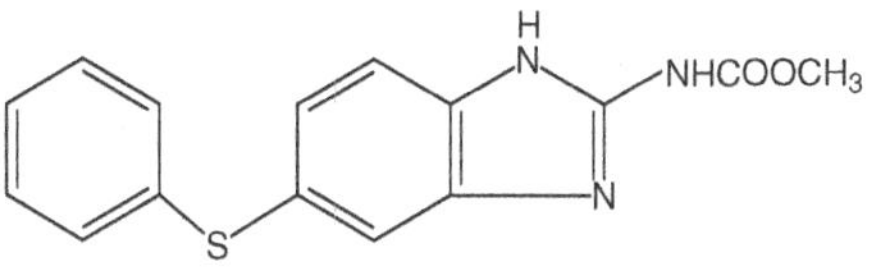

FENBENDAZOLE

subunit and the nucleotide is not hydrolyzable. The main function of the GTP on this site is to maintain the correct conformation of the polymer protein.

Antimicrotubule Drugs

Tubulin molecules bind a variety of different chemical agents. Colchicine is the oldest compound known and is the best studied. It is an alkaloid of the autumn crocus and has been used for the treatment of gout since ancient Egyptian times. When added to dividing cells each molecule binds firmly to one free tubulin molecule and prevents its polymerization. Colchicine cannot bind to tubulin once it becomes part of the polymer in the microtubule. Because of the inability of the free tubulin–colchicine complex of forming the microtubules, the process of polymerization to form functioning microtubules is blocked. One result of this effect is the arrest of dividing cells at the metaphase stage because fully polymerized microtubules are essential for moving chromosomes. Because of this disruption, colchicine and related drugs such as colcemid, vinblastine, and vincristine, which act in a similar way, are used against abnormally dividing cancerous cells.

Both biochemical and genetic studies have identified the β-tubulin subunit as the primary target for benzimidazole anthelmintics. Binding of these drugs to tubulin blocks the formation of microtubule matrix. Many of the biochemical processes that were originally thought to be primary sites of action of the anthelmintics are now thought to be affected only in a secondary way (Prichard, 1970; Van den Bossche & De Nollin, 1973). The effect of *in vivo* treatment with mebendazole on the ultrastructural morphology of *Ascaris suum* provided the first evidence that helminth microtubules are involved in the anthelmintic action of benzimidazoles against *Ascaris* (Borgers & De Nollin, 1975). Midgut cells of *Ascaris*, collected from infected pigs after 15–24 hours of treatment, showed acute degenerative changes that involved almost the entire cytoplasm and included malformation of microvilli located in the caecum of these parasites. This irreversible damage causes impaired uptake of nutrients, which results in starvation, death, and subsequent expulsion of the worms from the host.

It is now generally accepted that the mechanism of anthelmintic action of benzimidazoles is explained by their binding to tubulin and blocking its polymerization to form the microtubular matrix essential to the normal function of all eukaryotic cells. The question we now ask is: Why would a highly conserved protein such as tubulin, which is present in the host as well as the parasite, be the target for benzimidazoles? The answer lies in the selective binding of the benzimidazoles to nematode tubulin (Lacey & Gill, 1994). The binding of mebendazole or fenbendazole (two of the first benzimidazoles) to *A. suum* tubulin was 250 times greater than to mammalian tubulin. In addition, strains that showed clinical resistance to benzimidazoles were less able to bind the drug. Differential binding potency of the benzimidazoles, mebendazole, and fenbendazole to *A. suum* embryonic tubulin compared to bovine brain tubulin was reported (Friedman & Platzer, 1980a,b). The benzimidazoles bind to the colchicine binding domain and are effective inhibitors of colchicine binding to *A. suum* tubulin. For a review see Lacey (1990).

The relation between binding affinity and anthelmintic efficacy was examined by Lubega and Prichard on *Haemonchus contortus*. There is good correlation between binding constants for most of the benzimidazole anthelmintics to tubulin and their antiparasitic efficacy (Lubega & Prichard, 1991). All these experiments give more support to the idea that tubulin binding is involved in the mechanism of action of benzimidazoles. The mechanism of resistance to this group of anthelmintics involves a change in the ability of tubulin to bind the benzimidazole.

Benzimidazoles as Anthelmintics

Thiabendazole, 4-(2-benzimidazolyl)thiazole, a derivative of this series, became widely accepted against several species of human and animal gastrointestinal nematodes. Other benzimidazole compounds that were subsequently introduced include mebendazole, albendazole, and fenbendazole. For a review of the chemistry of this series see Fisher (1986). Although thiabendazole is effective against a variety of nematodes it is particularly active against infections by the nematode *Strongylides stercoralis*, which is resistant to the effect of other anthelmintics. Mebendazole is particularly effective against mixed infections of *Ascaris, Enterobius, Trichiuria,* and hookworm. Albendazole is superior to mebendazole against hookworm infections in humans (Tracy & Webster, 1995). The antiparasitic spectrum of benzimidazoles is broad and includes some cestodes and trematodes. Albendazole has been recommended against fascioliasis (Boray, 1986).

Resistance to Benzimidazoles

The effectiveness of benzimidazole anthelmintics against a wide spectrum of human and animal gastrointestinal parasites, added to the advantage of their low cost and ease of administration, led to their extensive use and, lately, the appearance of resistance. Research has been initiated to better understand the biology of the changes in sensitivity of the parasites to the benzimidazoles. Much of this research has been carried out in Australia because of the economic importance of the sheep industry and the spread of resistance to these anthelmintics. The infections studied are nematodes such as *Haemonchus contortus, Trichostronyglus colubriformis,* and *Ostertagia circumcincta.* For a review see Dobson, LeJambre, & Gill, 1996). Resistance to the benzimidazoles is inherited as an incomplete dominant/incomplete recessive trait. At least two genetic loci have been identified as being involved in expression of the resistance.

Genetic variability of the β-tubulin genes in benzimidazole-susceptible and -resistant strains of *H. Contortus* has been investigated. Results from Southern hybridization experiments identified two β-tubulin genes termed isotype-1 and isotype-2 (Roos, 1990; Beech, Prichard, & Scott, 1994; Kwa *et al.*, 1995). The number of alleles in susceptible strains was higher at both loci than in those of resistant strains. Experimental evidence indicates that development of benzimidazole resistance involves two molecular steps at the two separate β-tubulin loci. First there is an initial loss of susceptibility to the anthelmintic that results in reduction in alleles at the isotype-1 locus. This is followed by high benzimidazole resistance and allele deletion at the isotype-2 locus. Thus although both loci are involved in the response to resistance the changes in isotype-2 occur in the most resistant populations. A selection process is therefore involved to eliminate susceptible strains from those that are mildly or highly resistant strains. These findings led to the development of new assays to monitor benzimidazole-susceptible populations. Polymerase chain reaction (PCR) assays have been developed to monitor changes in allele frequency in benzimidazole resistance to parasitic infections of individual sheep. The same methods can be used to test parasite eggs and larvae in addition to the adults. Because benzimidazole resistance is heritable this new technology will make it possible to study the population genetics of sensitive and resistant parasites and different genetic parameters affecting selection and reversion with great accuracy.

Neuroactive Peptides in Nematodes

In mammalian neuropharmacology the search for peptides with neurotransmitter functions, either agonistic or antagonistic, has attracted considerable attention during the past 50 years. The number of peptides that affect the central and peripheral nervous system exceeds several dozen. Their selective functions in the nervous system in health and disease remains obscure. In the 1980s it was first established that the nematode nervous system contains a large number of peptides with inhibitory or excitatory effects on parasite mobility. Very few effects of peptides on specific neurons have been identified. Antisera raised to a variety of different vertebrate and invertebrate peptides have been used to identify by immunocytochemical means the presence of homologous peptides in *A. suum*, *Haemonchus contortus*, and *C. elegans*. Although the interest in nematode neuropeptides is only recent, a broad range of peptides have been identified. The invertebrate FMRFamide peptides (identified by single-letter amino acid symbols and commonly known by the acronyms FLPs or FaRPs) have been found by immunoreactivity in 75% of the nerve cells of *Ascaris*. Nineteen peptides structurally related to FMRFamide have been identified in *Ascaris* and one in *H. contortus* (Brownlee & Fairweather, 1999).

Although the selective function of these neuropeptides in the nematode nervous system remains to be understood, general effects on body wall muscle motility have been demonstrated. Both excitatory and inhibitory effects on *Ascaris* somatic muscle strip preparations have been attributed to some of the FMRFamide-related neuropeptides. The excitatory effect by KNEFIRFamide (AF1) in *Ascaris* is relaxation followed by increased muscle contractility. This increase in muscular activity at a micromolar concentration of the neuropeptide can extend for up to 2 hours (Brownlee *et al.*, 1993). The neuropeptide

SDPNFIRFamide (PF) causes slow Ca^{2+}-dependent flaccid paralysis of muscle strips at nanomolar concentrations. The endogenous nematode neuropeptide, KSAYMRFamide (AF8), is of special interest because at nanomolar concentrations it induces in the pharyngeal muscle an initial increase in amplitude and frequency of pumping followed by a reversible inhibition lasting up to 15 minutes (Brownlee *et al.*, 1995).

Searching for a new antiparasitic agent based on the current information on nematode neuropeptides has not been attempted. More research is needed, first, on the nature of the neuropeptide receptors; then, on the possible participation of signal transduction molecules such as G proteins; and finally on the possible cross-responses of several receptors to one neuropeptide. The dynamic interaction among different neuropeptide receptors as well as their interaction with the classical transmitters that regulate contractility (e.g., acetylcholine and GABA), should be studied before we can understand the effect of neuropeptides on the coordination of the nervous system functions. In addition, research on the possible effects of these neuropeptides on the mammalian nervous system is necessary. Turning on or off the neuropeptide receptors of the parasites by peptide congeners of naturally identified peptides requires more detailed information on their effect both on the parasite and on the mammalian host.

The Neuromuscular System of Flatworms

The liver fluke *Fasciola hepatica* and the blood fluke *Schistosoma mansoni* have been used as representatives of the trematodes for studies on motility. The musculature of these parasites is arranged predominantly in outer circular and inner longitudinal groups of smooth type muscle fibers. There is no evidence for the presence of striated muscle fibers. These parasites normally exhibit undulating body movements. In addition, both organisms have suckers composed of circular muscles by which they attach themselves to the host tissue during movement. The oral sucker participates in the process of food ingestion. The musculature of most of the tapeworms (cestodes) is similar to that of trematodes with the exception that the anterior segment (the scolex) has a sucker and some species have hooks. Cestodes do not have a digestive tract like trematodes. They absorb all their nutrients through their outer body layer. Metabolic products are also excreted through the outer body surface. The posterior segments of the cestodes (proglottids) contain primarily sexual organs. The central component of the nervous system in both trematodes and cestodes comprises paired cerebral ganglia located in the anterior end of the parasite. These supply the anterior and posterior main nerve trunks, which branch off to the rest of the body, including the suckers, the subtegument, and the peripheral motor and sensory plexuses. The parasite peripheral neuromuscular system also controls organs inside the parasite, including the caecum, pharynx (in trematodes), and genitalia (necessary for egg laying) (Smyth, 1994).

Evaluation of Motility of Flatworms

Because many flatworms can be eliminated from the body if they cease to move, the search for anthelmintics that interfere with motility necessitated an accurate measure of muscular activity. Early studies on the motility of flatworms were carried out on intact parasites kept *in vitro* in suitable media. Subsequently, physiological and pharmacological *in vitro* studies on motility used either intact or degangliated preparations or neuromuscular preparations (a strip of the parasite body) suspended in organ baths. Because of its large size and availability, *Fasciola hepatica* were used for early studies. The changes in rate, amplitude, and tone of contractions of these preparations could be measured using a kymograph or strain gauge. The effects of neuromuscular drugs that inhibit motility were studied in the search for possible anthelmintics (Chance & Mansour, 1949, 1953). Recently, elaborate procedures using multiphotocells were devised to measure motility of parasites such as *S. mansoni* that were too small for making organ bath preparations (Fetterer, Pax, & Bennett, 1977). In cestodes, physiological and pharmacological experiments have been carried out on preparations similar to those described above for *Fasciola hepatica* (Sano *et al.*, 1982; Terada *et al.*, 1982).

5-Hydroxytryptamine (Serotonin) in Parasitic Flatworms

Serotonin appears to function in both trematodes and cestodes as an excitatory neurotransmitter regulating motility of these parasites. This effect has been demonstrated in the trematodes *F. hepatica, S. mansoni,* and *Clonorchis sinensis* and the cestodes *Hymenolepis diminuta* and *Dipelidium caninum* (Mansour, 1964; Barker, Bueding, & Timms, 1966; Mettrick & Cho, 1981; Terada *et al.*, 1982). Furthermore, biochemical effects similar to those found in *F. hepatica* have also been demonstrated in a variety of both trematodes and cestodes. For references see reviews by Mansour (1984) and Pax *et al.* (1996). Various agonists and antagonists of serotonin were examined for their effects on motility in intact parasites and on adenylate cyclase activity and binding to serotonin receptors in parasite homogenates (McNall & Mansour, 1984; Estey & Mansour, 1987). The results showed strong correlations between these effects, supporting the view that there may be similarities in the mechanism of action. Serotonin also appears to act as a hormone-like activator of the carbohydrate metabolism of these parasites (see Chapter 3).

Early experiments with *Fasciola hepatica* showed that serotonin stimulates rhythmical movement of the parasite. It is superior to other indolamines such as tryptamine, bufotenine (N, N-dimethyl serotonin), and 1-benzyl-5-methoxy-2-methyl tryptamine in stimulating rhythmical movement (Mansour, 1957). Other bioamines, such as the catecholamines, epinephrine, and norepinephrine, show no effects similar to those of serotonin.

Serotonin was identified by histochemical, biochemical, and immunocyto-chemical methods and bioassay in a variety of trematode and cestode species. Serotonin concentrations were more pronounced in different parts of the central nervous system including the anterior ganglia and the longitudinal and transverse nerve cords. Serotonin was also identified in peripheral nerve fibers associated with subtegumental muscles, the sucker muscles, and the muscular linings of various reproductive ducts. For a review see Davis & Stretton (1995).

As in mammalian cells, the biosynthesis of serotonin from tryptophan in trematodes depends on the participation of two enzymes: First tryptophan hydroxylase (TPH) catalyzes the hydroxylation of tryptophan to 5-hydroxytryptophan (5-HTP); then 5-hydroxytryptophan decarboxylase catalyzes the decarboxylation of 5-HTP to 5-hydroxytryptamine (serotonin). The presence of both enzyme systems has been established in both *F. hepatica* and *S. mansoni* (Mansour & Stone, 1970; Bennett & Bueding, 1973; Hamdan & Ribeiro, 1999).

The identification of the synthetic enzymes of serotonin in representative trematodes shows that these parasites can synthesize serotonin from tryptophan using their own enzymes. Serotonin is also available to the parasite from outside sources. It is present in the blood serum, the platelets, and the mast cells of the host. *In vitro* studies on *S. mansoni* showed that, at low serotonin concentrations (1–5 μM) in the host, there is an uptake mechanism for serotonin that has saturable kinetics. This appears to be the physiological route for serotonin uptake by the parasites from the circulating blood of the host (Bennett & Bueding, 1973; Wood & Mansour, 1986). The availability of serotonin from both sources may satisfy the special regulatory needs of the parasites during their development. In addition to its known effect on neuromuscular activity and carbohydrate metabolism, studies with other invertebrates have identified serotonin as a modulator of feeding and egg release. It is probable that other aspects of behavior that are modified by serotonin have yet to be discovered (Weiger, 1997).

Cholinergic Systems in Flatworms

Acetylcholine and the enzymes necessary for its synthesis (choline acetyltransferase) and degradation (acetylcholinesterase) have been identified in many trematodes and cestodes. Pharmacological experiments were carried out to identify the type of cholinergic receptors present in these organisms. In *Fasciola hepatica* acetylcholine, its congeners (e.g., carbachol), and acetylcholinesterase inhibitors (e.g., physostigmine) relax the musculature of the worms, inhibit rhythmical movement, and eventually produce paralysis (Chance & Mansour, 1949, 1953). These effects appear to be associated with the peripheral nervous system as parasite preparations that have had their central ganglia removed show the same responses. Similar decreases in motility have been demonstrated in the cestode *Dipylidium caninum* (Terada *et al.*, 1982) and other cestodes. Further characterization of these cholinergic responses as muscarinic or nicotinic types was not achieved: Both types of effects were seen. This is probably because the classical antagonists used in these experiments were chosen for their selective effects on vertebrate cholinergic receptors. The parasite receptors may be different. In addition, the pharmacological preparations usually used for these studies may contain multiple pathways, representative of both muscarinic and nicotinic receptors (Davis & Stretton, 1995).

Influence of Parasite Motility on Parasite Location in the Host

Parasite movement appears to be influenced by serotonin. For example, migration behavior of the tapeworm *Hymenolepis diminuta* in the small intestine of the rat (movement toward the anterior end of the intestine) is influenced by circadian changes in the parasites. These are correlated with host feeding, followed by a rise in serotonin levels in both host tissues and worms. Furthermore, serotonin administered to the host parenterally or by mouth causes the parasites to migrate up the intestine (Mettrick & Cho, 1981; Cho & Mettrick, 1982).

Fasciola hepatica depend to a large extent on the use of their suckers and coordination of their movement for maintaining their attachment to the wall of the host's bile duct. A prime effect of mebendazole against this infection in sheep is parasite detachment from the bile duct. Treatment with rafoxanide (a diiodosalicylanilide) of sheep infected with *Fasciola hepatica* also causes the parasites to detach from the bile duct walls. This results in cessation of feeding and the flukes are carried by the flow of the bile to the gall bladder (a much more acidic milieu than the bile duct) and ultimately to the intestine. As the flukes are expelled from their natural habitat in the bile duct many die on the way before being expelled with the intestinal contents. One important consequence of their initial detachment from the bile ducts is the progressive deterioration of their energy status because of their inability to feed from the host's hepatic venules (Cornish *et al.*, 1977; Rahman *et al.*, 1977; Chevis, 1980). This *in vivo* effect of a fasciolicide on motility has been corroborated by *in vitro* studies of contractility of *F. hepatica* recorded by means of an isometric transducer system. Mebendazole causes suppression of motility in these parasites (Fairweather, Holmes, & Threadgold, 1984).

One way of eliminating different parasites from the gastrointestinal tract is to paralyze them so that their suckers detach from the host. The incapacitated parasites can then be swept away with the gut contents. Presumably, incoordination of their motility and loss of sucker control could result in dislocation from other sites as well. Among parasites that live in the blood vessels are *Schistosoma mansoni* and immature stages of *Fasciola hepatica*. Changes in the motility of these parasites have been shown to cause a change in their localization. In mice infected with *S. mansoni* the majority (74%) of adult parasites are located in the mesenteric venules whereas 21% are in the main portal vein and only 4% reside in the liver venules. Within half an hour after treatment with the schistosomicidal drug antimony potassium tartrate almost all the adults are found in the liver. Continuing treatment with low doses of the drug may not kill the parasites. They regain their motility and move back to the mesenteric venules. This rapid change in the position of the worms from the mesentery to the liver is known as the "hepatic shift." It has been assumed that migration of the worms from the mesentery to the liver primarily occurs because of the loss of their muscle tone and their grip in the mesenteric venules. As a result the worms are swept back into the liver by the flow of blood. Worm migration back from the liver to the mesenteric venules is slower because they will be moving against the bloodstream (Buttle & Khayyal, 1962). When the antischistosomal agent is given in a therapeutic dose that forces the parasites to remain in the liver venules for a sufficiently long time, the parasites become immobilized in the narrow liver venules and as a result cease to feed. The presence of the parasites causes the walls of the hepatic vessels to become inflamed. Fibrosis then occurs and finally the vessels are occluded. Eventually, the parasites die and their tissues are invaded by leukocytes (Standen, 1953). The migration of schistosomes from the mesentery to the liver has been used as a method for identifying drugs that have schistosomicidal effects (Standen, 1953).

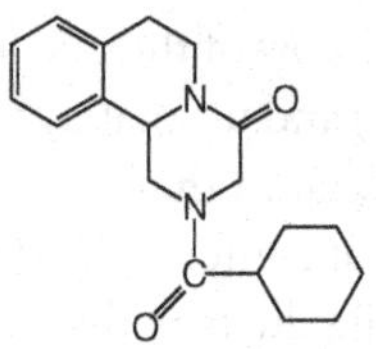

PRAZIQUANTEL

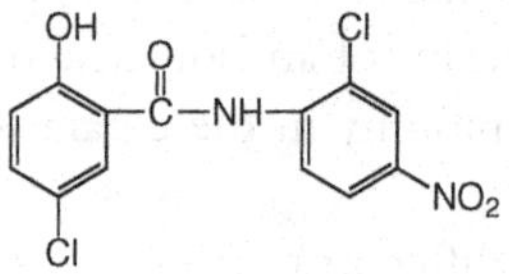

NICLOSAMIDE

Anthelmintics That Act on Motility in Platyhelminths

Because parasites are dependent for their survival on living in a specific site in the host, drugs that interfere with parasite motility could initiate their removal from the host. For example, praziquantel, a drug that has been used with some success against *Fasciola hepatica* as well as schistosome infections in humans, affects motility (Fairweather *et al.*, 1984). Neuromuscular preparations of *F. hepatica* show spastic paralysis when praziquantel is added to the organ bath for recording motility. Similarly, when motility of *Schistosoma mansoni* was recorded, praziquantel addition causes a rapid increase in tension followed by spastic paralysis (Fetterer, Pax, & Bennett, 1980). In cestodes both praziquantel and niclosamide are considered the drugs of choice against human infection by *Taenia, Diphyllobothrium,* and *Hymenolepis. In vitro* tests of these anthelmintics on preparations of these parasites cause rapid spastic paralysis. This effect appears to be the prime cause of the parasite's elimination from the host (Sano *et al.*, 1982; Terada *et al.*, 1982). More information on these drugs is presented in Chapter 8.

Neuropeptides in Flatworms

Although the major research on neuropeptides in parasites has been done on nematodes, recent reports indicate that neuropeptides could play a role in the control of neuromuscular function in platyhelminths. An FaRP-related peptide with the sequence GNFFRFamide was identified in the nervous system of the sheep tapeworm *Moniezia expansa* (Maule *et al.*, 1993). The same peptide was identified in the trematodes *Fasciola hepatica* and *Schistosoma mansoni* (Marks *et al.*, 1995). FMRFamide-related peptides are said to have a myoactive effect when applied to trematode muscle (Pax *et al.*, 1996). As can be seen, more studies are needed regarding characterization of these peptides in platyhelminths and the nature of their pharmacological action on the neuromuscular system before the receptors can be used as targets for prospective antiparasitic agents.

Potential Research on Anthelmintics

During the past quarter century there has been amazing progress in understanding the biology of the mammalian neuromuscular system. This included the structure and function of neurons, interneuronal signaling, and the role played by neurotransmitter receptors. Our current understanding of the biology of the neuromuscular system in parasitic helminths is in its beginnings. Much of our information came about as a result of the discovery of a few anthelmintic drugs that happen to affect parasite motility. This favorable development stimulated basic investigations on the neuroanatomy, physiology, and pharmacology of model parasitic helminths, such as *Ascaris* and *Fasciola hepatica*. It established the beginning of electrophysiological experimentation with neuromuscular preparations of the parasites. The expectation is that more information will soon become available on the physiology and pharmacology of different species of parasitic worms. One flaw in current experimental strategies is the entire dependence on methodologies acquired from neuromuscular investigations in mammals. Observers have tended to disregard the fact that obligate parasites' motility in their natural host habitats could be different from that in laboratory media. Parasites are known to be influenced by changes in their milieu. Already mentioned are changes in the migration behavior of tapeworms influenced by circadian variations in their serotonin levels caused by host feeding. Studies on the nature of parasite motility should be done either in the mammalian host or in media closely related to the parasites' natural environment. It would also be appropriate to develop methods for measuring the movement of organelles in the parasites such as the nematode pharynx (Geary *et al.*, 1993) or oviduct. Some neurotransmitters have been found to influence motility of the pharynx (which is necessary for feeding) at much lower concentrations than those affecting somatic movement.

The availability of the complete sequence of *C. elegans* will have a profound impact on the search for novel anthelmintics. It is now possible to identify homologous genes that are critical for survival of the parasite and use them as probes for the isolation of parasite genes from cDNA libraries. *C. elegans* genes are also being used for preparing knockout mutants to verify the importance of an anthelmintic target for the survival of the parasite. The use of the *C. elegans* genome sequence in techniques for reverse genetics, screening for lethal mutants, and rescue transformation experiments provide other powerful physical and genetic tools for identifying putative drug targets. For a review see Thompson, Klein, & Geary (1996). In spite of the great importance of the role that *C. elegans* plays and will continue to play in genetic and molecular research, the free-living nematode cannot be used as a replacement for parasitic helminths in physiological and pharmacological experimentation in the search for new anthelmintics.

Clones of genes from *C. elegans* have been used to isolate nicotinic acetylcholine receptor subunits from the human parasitic nematode *Onchocerca volvulus* (a filaria) and the sheep nematode *Trichostrongylus colubriformis* (Fleming *et al.*, 1996). It is difficult to obtain enough *O. volvulus* parasites for molecular or biochemical studies. This is a great handicap for studying parasite receptors. The nature of the cloned receptors or the effect of chemical agents on their function can be further examined by expression of the receptor gene in *Xenopus* oocytes. Functional studies on these expressed genes could provide more information on the mode of action of known anthelmintics and the mechanism of resistance. The receptors could be used as a target for selecting new and more effective anthelmintics. The only standard nicotinic anthelmintic group that is commonly used for sheep and pigs is the levamisole/pyrantel group. Nematode infections in these animals are showing signs of significant resistance to these anthelmintics (Sangster, 1996). There is a need for a new group of nicotinic anthelmintics.

GABA receptors are another component of the nervous system of nematode parasites that deserve special investigation for identifying new anthelmintic agents. The only selective agonist of the GABA-gated chloride channels in *Ascaris* muscle membrane is piperazine, which causes paralysis of the worms (Martin *et al.*, 1997). Although piperazine was at one time the drug of choice against pinworm infections in children it has long been abandoned because of some side effects in patients. The gene *unc-49* was shown in *C. elegans* to be necessary for the inhibitory effect of GABA on body muscles. The gene appears to encode for a GABA receptor subunit that controls the inhibitory effect of the neurotransmitter on the parasite muscles (Fleming *et al.*, 1996). A cDNA encoding another inhibitory amino acid receptor that shows homology with invertebrate GABA-gated Cl^- channels has been cloned from the parasitic nematode *Haemonchus contortus* (Laughton *et al.*, 1994). Establishing functional expression of *C. elegans* and *H. contortus* GABA receptors in *Xenopus* oocytes would be a great contribution toward better understanding of these receptors and for the selection of new anthelmintics more potent than piperazine. Critical questions that need to be answered are: 1. What are the interactions between different subunits of the receptor? and 2. What are the functions of the different subunits?

Both nematode and flatworm parasites as well as *C. elegans* have serotonin receptors that regulate neuromuscular activity. In addition to somatic movement, pharyngeal pumping, which is essential for nematode feeding and survival, appears to be controlled by serotonin receptors. These serotonin receptors are considered attractive targets for interfering with parasitic motility. Serotonin analogs, such as bromo-LSD, bind to the receptors and antagonize their coupling with adenylate cyclase, causing parasite paralysis (see Chapter 3). The availability of serotonin receptor clones and the recombinant proteins should facilitate screening almost any number of serotonin agonists and antagonists. Most of the serotonin analogs that have already been tested on parasites have been drawn from programs to select drugs interacting with human serotonin receptors. Now, with the parasite receptors available, it will become possible to identify highly selective inhibitors.

REFERENCES

Abramson, S. N., Culver, P., Kline, T., Li, Y., Guest, P., Gutman, L. & Taylor, P. (1988). Lophotoxin and related coral toxins covalently label the alpha-subunit of the nicotinic acetylcholine receptor. *J Biol Chem, 263*(34), 18568–18573.

Abramson, S. N., Li, Y., Culver, P. & Taylor, P. (1989). An analog of lophotoxin reacts covalently with Tyr190 in the alpha-subunit of the nicotinic acetylcholine receptor. *J Biol Chem, 264*(21), 12666–12672.

Aceves, J., Erlij, D. & Martinez-Maranon, R. (1970). The mechanism of the paralysing action of tetramisole on *Ascaris* somatic muscle. *Br J Pharmacol, 38*(3), 602–607.

Ajuh, P. M. & Egwang, T. G. (1994). Cloning of a cDNA encoding a putative nicotinic acetylcholine receptor subunit of the human filarial parasite *Onchocerca volvulus*. *Gene, 144*(1), 127–129.

Alberts, B., Bray, D., Lewis, J., Raff, M., Roberts, K. & Watson, J. D. (1994). *Molecular Biology of the Cell* (3rd ed.). New York: Garland Publishing.

Arena, J. P., Liu, K. K., Paress, P. S., Frazier, E. G., Cully, D. F., Mrozik, H. & Schaeffer, J. M. (1995). The mechanism of action of avermectins in *Caenorhabditis elegans*: Correlation between activation of glutamate-sensitive chloride current, membrane binding, and biological activity. *J Parasitol, 81*(2), 286–294.

Avery, L. (1993). Motor neuron M3 controls pharyngeal muscle relaxation timing in *Caenorhabditis elegans*. *J Exp Biol, 175*, 283–297.

Avery, L. & Horvitz, H. R. (1990). Effects of starvation and neuroactive drugs on feeding in *Caenorhabditis elegans*. *J Exp Zool, 253*(3), 263–270.

Baldwin, E. (1943). An *in vitro* method for the chemotherapeutic investigation of anthelmintic potency. *Parasitology, 35*, 89–111.

Baldwin, E. & Moyle, V. (1949). A contribution to the physiology and pharmacology of *Ascaris lumbricoides* from the pig. *Br J Pharmacol, 4*, 145–152.

Barker, L. R., Bueding, E. & Timms, A. R. (1966). The possible role of acetylcholine in *Schistosoma mansoni*. *Br J Pharmacol, 26*(3), 656–665.

Beech, R. N., Prichard, R. K. & Scott, M. E. (1994). Genetic variability of the beta-tubulin genes in benzimidazole-susceptible and -resistant strains of *Haemonchus contortus*. *Genetics, 138*(1), 103–110.

Bennett, J. L. & Bueding, E. (1973). Uptake of 5-hydroxytryptamine by *Schistosoma mansoni*. *Mol Pharmacol, 9*(3), 311–319.

Boray, J. (1986). Trematode infections of domestic animals. In W. Campbell & R. Rew (Eds.), *Chemotherapy of Parasitic Diseases* (pp. 401–425). New York: Plenum Press.

Borgers, M. & De Nollin, S. (1975). Ultrastructural changes in *Ascaris suum* intestine after mebendazole treatment in vivo. *J Parasitol, 61*(1), 110–122.

Broeks, A., Janssen, H. W., Calafat, J. & Plasterk, R. H. (1995). A P-glycoprotein protects *Caenorhabditis elegans* against natural toxins. *EMBO J, 14*(9), 1858–1866.

Brown, H. D., Matzuk, A. R., Ilves, I. R., Perterson, L. H., Harris, S. A., Sarett, L. H., Egerton, J. R., Yakstis, J. J., Campbell, W. C. & Cuckler, A. C. (1961). Antiparasitic drugs. IV. 2-(4'-thiazolyl)-benzimidazole, a new anthelmintic. *J Am Chem Soc, 83*, 1764–1765.

Brownlee, D. J. & Fairweather, I. (1999). Exploring the neurotransmitter labyrinth in nematodes. *Trends Neurosci, 22*(1), 16–24.

Brownlee, D. J., Fairweather, I., Johnston, C. F., Smart, D., Shaw, C. & Halton, D. W. (1993). Immunocytochemical demonstration of neuropeptides in the central nervous system of the roundworm, *Ascaris suum* (*Nematoda: Ascaroidea*). *Parasitology, 106*(Pt 3), 305–316.

Brownlee, D. J., Holden-Dye, L., Fairweather, I. & Walker, R. J. (1995). The action of serotonin and the nematode neuropeptide KSAYMRFamide on the pharyngeal muscle of the parasitic nematode, *Ascaris suum*. *Parasitology, 111*(Pt 3), 379–384.

Brownlee, D. J., Holden-Dye, L. & Walker, R. J. (1997). Actions of the anthelmintic ivermectin on the pharyngeal muscle of the parasitic nematode, *Ascaris suum. Parasitology, 115*(Pt 5), 553–561.

Buttle, G. & Khayyal, M. (1962). Rapid hepatic shift of worms in mice infected with *Schistosoma mansoni* after a single injection of tartar emetic. *Nature, 194*, 780–781.

Campbell, W. C. (1985). Ivermectin: An update. *Parasitol Today, 1*, 10–16.

Campbell, W. C., Burg, R. W., Fisher, M. H. & Dybas, R. A. (1984). *Pesticide Synthesis through Rational Approaches* (Vol. 255). New York: Plenum Press.

Chance, M. R. A. & Mansour, T. E. (1949). A kymographic study of the action of drugs on the liver fluke (*Fasciola hepatica*). *Br. J. Pharmacol, 4*, 7–13.

Chance, M. R. A. & Mansour, T. E. (1953). A contribution to the pharmacology of movement in the liver fluke. *Br. J. Pharmacol, 8*, 134–138.

Chevis, R. A. (1980). The speed of action of anthelmintics: Mebendazole. *Vet Rec, 107*(17), 398–399.

Cho, C. H. & Mettrick, D. F. (1982). Circadian variation in the distribution of *Hymenolepis diminuta* (Cestoda) and 5-hydroxytryptamine levels in the gastro-intestinal tract of the laboratory rat. *Parasitology, 84*(Pt 3), 431–441.

Cornish, R. A., Behm, C. A., Butler, R. W. & Bryant, C. (1977). The in vivo effects of rafoxanide on the energy metabolism of *Fasciola hepatica. Int J Parasitol, 7*(3), 217–220.

Cully, D. F., Vassilatis, D. K., Liu, K. K., Paress, P. S., Van der Ploeg, L. H., Schaeffer, J. M. & Arena, J. P. (1994). Cloning of an avermectin-sensitive glutamate-gated chloride channel from *Caenorhabditis elegans. Nature, 371*(6499), 707–711.

Cully, D. F., Paress, P. S., Liu, K. K., Schaeffer, J. M. & Arena, J. P. (1996a). Identification of a *Drosophila melanogaster* glutamate-gated chloride channel sensitive to the antiparasitic agent avermectin. *J Biol Chem, 271*(33), 20187–20191.

Cully, D. F., Wilkinson, H., Vassilatis, D. K., Etter, A. & Arena, J. P. (1996b). Molecular biology and electrophysiology of glutamate-gated chloride channels of invertebrates. *Parasitology, 113* (Suppl), S191–S200.

Davis, R. E. (1998). Neurophysiology of glutamatergic signalling and anthelmintic action in *Ascaris suum*: Pharmacological evidence for a kainate receptor. *Parasitology, 116*(Pt 5), 471–486.

Davis, R. E. & Stretton, A. O. W. (1995). Neurotransmitters of Helminths. In J. J. Marr & M. Muller (Eds.), *Biochemistry and Molecular Biology of Parasites* (pp. 257–288). San Diego: Academic Press.

Davis, R. E. & Stretton, A. O. (1996). The motornervous system of *Ascaris*: Electrophysiology and anatomy of the neurons and their control by neuromodulators. *Parasitology, 113*(Suppl), S97–117.

Del Castillo, J., De Mello, W. & Morales, T. (1963). The physiologial role of acetylcholine in the neuromuscular system of *Ascaris lumbricoides*. *Arch Int Physiol Biochim, 71*, 741–757.

Del Castillo, J., De Mello, W. C. & Morales, T. (1964). Inhibitory action of gammaaminobutyric acid (GABA) on *Ascaris* muscle. *Experientia, 20*(3), 141–143.

Dobson, R. J., LeJambre, L. & Gill, J. H. (1996). Management of anthelmintic resistance: Inheritance of resistance and selection with persistent drugs. *Int J Parasitol, 26*(8–9), 993–1000.

Erspamer, V. (1954). Pharmacology of indolamine alkylamines. *Pharmacol Rev, 6*, 425–487.

Estey, S. J. & Mansour, T. E. (1987). Nature of serotonin-activated adenylate cyclase during development of *Schistosoma mansoni*. *Mol Biochem Parasitol, 26*(1–2), 47–59.

Fairweather, I., Holmes, S. D. & Threadgold, L. T. (1984). Fasciola *hepatica*: Motility response to fasciolicides in vitro. *Exp Parasitol, 57*(3), 209–224.

Fetterer, R. H., Pax, R. A. & Bennett, J. L. (1977). *Schistosoma mansoni*: Direct method for simultaneous recording of electrical and motor activity. *Exp Parasitol, 43*(1), 286–294.

Fetterer, R. H., Pax, R. A. & Bennett, J. L. (1980). Praziquantel, potassium and 2,4-dinitrophenol: Analysis of their action on the musculature of *Schistosoma mansoni*. *Eur J Pharmacol, 64*(1), 31–38.

Fisher, M. H. (1986). Chemsitry of antinematodal agents. In C. C. Campbell & R. S. Rew (Eds.), *Chemotherapy of Parasitic Deiseases* (pp. 239–264). New York: Plenum Press.

Fleming, J. T., Baylis, H. A., Sattelle, D. B. & Lewis, J. A. (1996). Molecular cloning and in vitro expression of *C. elegans* and parasitic nematode ionotropic receptors. *Parasitology, 113*(Suppl), S175–190.

Friedman, P. A. & Platzer, E. G. (1980a). Interaction of anthelmintic benzimidazoles with *Ascaris suum* embryonic tubulin. *Biochim Biophys Acta, 630*(2), 271–278.

Friedman, P. A. & Platzer, E. G. (1980b). *The molecular mechanism of action of benzimaidazoles in embyros of Ascaris suum*. Paper presented at the 3rd International Symposium on: The Biochemistry of Parasites and Host-Parasite Relationships, Beerse, Belgium.

Geary, T. G., Sims, S. M., Thomas, E. M., Vanover, L., Davis, J. P., Winterrowd, C. A., Klein, R. D., Ho, N. F. & Thompson, D. P. (1993). *Haemonchus contortus*: Ivermectin-induced paralysis of the pharynx. *Exp Parasitol, 77*(1), 88–96.

Greene, B. M., Taylor, H. R., Cupp, E. W., Murphy, R. P., White, A. T., Aziz, M. A., Schulz-Key, H., D'Anna, S. A., Newland, H. S., Goldschmidt, L. P. *et al.* (1985). Comparison of ivermectin and diethylcarbamazine in the treatment of onchocerciasis. *N Engl J Med, 313*(3), 133–138.

Guastella, J. & Stretton, A. O. W. (1991). Distribution of H² GABA uptake sites in the nematode *Ascaris. J Comp Neurol, 307*, 598–608.

Hamdan, F. F. & Ribeiro, P. (1999). Characterization of a stable form of tryptophan hydroxylase from the human parasite *Schistosoma mansoni. J Biol Chem, 274*(31), 21746–21754.

Horvitz, H. R., Chalfie, M., Trent, C., Sulston, J. E. & Evans, P. D. (1982). Serotonin and octopamine in the nematode *Caenorhabditis elegans. Science, 216*(4549), 1012–1014.

Huang, X., Duran, E., Diaz, F., Xiao, H., Messer, W. S., Jr. & Komuniecki, R. (1999). Alternative-splicing of serotonin receptor isoforms in the pharynx and muscle of the parasitic nematode, *Ascaris suum. Mol Biochem Parasitol, 101*(1–2), 95–106.

Johnson, C. & Stretton, A. (1980). Neural control of locomotion in *Ascaris*: Anatomy, electrophysiology and biochemistry, In B. M. Zuckerman (Ed.), *Nematodes as Biological Models* (Vol. I). New York: Academic Press.

Kao, P. N. & Karlin, A. (1986). Acetylcholine receptor binding site contains a disulfide cross-link between adjacent half-cystinyl residues. *J Biol Chem, 261*(18), 8085–8088.

Karlin, A. & Akabas, M. H. (1995). Toward a structural basis for the function of nicotinic acetylcholine receptors and their cousins. *Neuron, 15*(6), 1231–1244.

Kwa, M. S., Veenstra, J. G., Van Dijk, M. & Roos, M. H. (1995). Beta-tubulin genes from the parasitic nematode *Haemonchus contortus* modulate drug resistance in *Caenorhabditis elegans. J Mol Biol, 246*(4), 500–510.

Lacey, E. (1990). Mode of action of benzimidazoles. *Parasitol Today, 6*, 112–115.

Lacey, E. & Gill, J. H. (1994). Biochemistry of benzimidazole resistance. *Acta Trop, 56*(2–3), 245–262.

Laughton, D. L., Amar, M., Thomas, P., Towner, P., Harris, P., Lunt, G. G. & Wolstenholme, A. J. (1994). Cloning of a putative inhibitory amino acid receptor subunit from the parasitic nematode *Haemonchus contortus. Receptors Channels, 2*(2), 155–163.

Lewis, J. A., Wu, C. H., Berg, H. & Levine, J. H. (1980a). The genetics of levamisole resistance in the nematode *Caenorhabditis elegans. Genetics, 95*(4), 905–928.

Lewis, J. A., Wu, C. H., Levine, J. H. & Berg, H. (1980b). Levamisole-resistant mutants of the nematode *Caenorhabditis elegans* appear to lack pharmacological acetylcholine receptors. *Neuroscience, 5*(6), 967–989.

Lubega, G. W. & Prichard, R. K. (1991). Interaction of benzimidazole anthelmintics with *Haemonchus contortus* tubulin: Binding affinity and anthelmintic efficacy. *Exp Parasitol, 73*(2), 203–213.

Mansour, T. & Stone, D. (1970). Biochemical effects of lysergic acid diethylamide on the liver fluke *Fasciola hepatica. Biochem Pharmacol, 19*, 1137–1145.

Mansour, T. E. (1957). The effect of lysergic acid diethylamide, 5-hydroxytryptamine, and related compounds on the liver fluke *Fasciola hepatica. Br J Pharmacol, 12*, 406–409.

Mansour, T. E. (1964). *The Pharmacology and Biochemistry of Parasitic Helminths* (Vol. 3). New York: Academic Press.

Mansour, T. E. (1984). Serotonin receptors in parasitic worms. *Adv Parasitol, 23*, 1–6.

Mansour, T. E., Lago, A. D. & Hawkins, J. L. (1957). Occurence and possible role of serotonin in *Fasciola hepatica. Fed Proc, 16*, 319.

Maricq, A. V., Peckol, E., Driscoll, M. & Bargmann, C. I. (1995). Mechanosensory signalling in *C. elegans* mediated by the GLR-1 glutamate receptor. *Nature, 378*(6552), 78–81. [Published erratum appears in Nature 1996, *379*(6567), 749.]

Marks, N. J., Halton, D. W., Maule, A. G., Brennan, G. P., Shaw, C., Southgate, V. R.

& Johnston, C. F. (1995). Comparative analyses of the neuropeptide F (NPF)- and FMRFamide-related peptide (FaRP)-immunoreactivities in *Fasciola hepatica* and *Schistosoma spp. Parasitology, 110*(Pt 4), 371–381.

Martin, R. J. (1982). Electrophysiological effects of piperazine and diethylcarbamazine on *Ascaris suum* somatic muscle. *Br J Pharmacol, 77*(2), 255–265.

Martin, R. J. (1985). Gamma-Aminobutyric acid- and piperazine-activated single-channel currents from *Ascaris suum* body muscle. *Br J Pharmacol, 84*(2), 445–461.

Martin, R. J. (1996). An electrophysiological preparation of *Ascaris suum* pharyngeal muscle reveals a glutamate-gated chloride channel sensitive to the avermectin analogue, milbemycin D. *Parasitology, 112*(Pt 2), 247–252.

Martin, R. J., Robertson, A. P. & Bjorn, H. (1997). Target sites of anthelmintics. *Parasitology, 114*(Suppl), S111–124.

Martin, R. J., Valkanov, M. A., Dale, V. M. E., Robertson, A. P. & Murray, I. (1996). Electrophysiology of *Ascaris* muscle and anti-nematodal drug action. *Parasitology, 113*(Suppl), S137–156.

Maule, A. G., Halton, D. W., Shaw, C. & Johnston, C. F. (1993). The cholinergic, serotoninergic and peptidergic components of the nervous system of *Moniezia expansa* (Cestoda, Cyclophyllidea). *Parasitology, 106*(Pt 4), 429–440.

McNall, S. J. & Mansour, T. E. (1984). Desensitization of serotonin-stimulated adenylate cyclase in the liver fluke *Fasciola hepatica. Biochem Pharmacol, 33*(17), 2799–2805.

Mettrick, D. F. & Cho, C. H. (1981). Migration of *Hymenolepis diminuta* (Cestoda) and changes in 5-HT (serotonin) levels in the rat host following parenteral and oral 5-HT administration. *Can J Physiol Pharmacol, 59*(3), 281–286.

Norton, S. & DeBeer, E. J. (1957). Investigations on the action of piperazine on *Ascaris lumbricoides. Am J Trop Med, 6*, 889–905.

Olde, B. & McCombie, W. R. (1997). Molecular cloning and functional expression of a serotonin receptor from *Caenorhabditis elegans. J Mol Neurosci, 8*(1), 53–62.

Pax, R. A., Day, T. A., Miller, C. L. & Bennett, J. L. (1996). Neuromuscular physiology and pharmacology of parasitic flatworms. *Parasitology, 113*(Suppl), S83–96.

Prichard, R. K. (1970). Mode of action of the anthelminthic thiabendazole in *Haemonchus contortus. Nature, 228*(272), 684–685.

Rahman, M. S., Cornish, R. A., Chevis, R. A. & Bryant, C. (1977). Metabolic changes in some helminths from sheep treated with mebendazole. *N Z Vet J, 25*(4), 79–83.

Richards, J. C., Behnke, J. M. & Duce, I. R. (1995). In vitro studies on the relative sensitivity to ivermectin of *Necator americanus* and *Ancylostoma ceylanicum. Int J Parasitol, 25*(10), 1185–1191.

Robertson, A. P., Bjorn, H. E. & Martin, R. J. (1999). Resistance to levamisole resolved at the single-channel level. *FASEB J, 13*(6), 749–760.

Rohrer, S. P., Birzin, E. T., Eary, C. H., Schaeffer, J. M. & Shoop, W. L. (1994). Ivermectin binding sites in sensitive and resistant *Haemonchus contortus. J Parasitol, 80*(3), 493–497.

Roos, M. (1990). The molecular nature of benzimidazole resistance in helminths. *Parasitol Today, 6*, 125–127.

Sangster, N. (1996). Pharmacology of anthelmintic resistance. *Parasitology, 113*, S201–216.

Sangster, N. C., Bannan, S. C., Weiss, A. S., Nulf, S. C., Klein, R. D. & Geary, T. G. (1999). *Haemonchus contortus*: Sequence heterogeneity of internucleotide binding domains from P-glycoproteins. *Exp Parasitol, 91*(3), 250–257.

Sano, M., Terada, M., Ishii, A. I., Kino, H. & Anantaphruti, M. (1982). Studies on chemotherapy of parasitic helminths (V). Effects of niclosamide on the motility of various parasitic helminths. *Experientia, 38*(5), 547–549.

Segalat, L., Elkes, D. A. & Kaplan, J. M. (1995). Modulation of serotonin-controlled behaviors by Go in *Caenorhabditis elegans. Science, 267*(5204), 1648–1651. [See comments.]

Smyth, J. (1994). *Introduction to Animal Parasitology* (3rd ed.). Cambridge: Cambridge University Press.

Standen, O. (1953). Experimental schistosomiasis III. – Chemotherapy and mode of drug action. *Ann Trop Med Parasitol, 47*, 26–43.

Stretton, A. O., Fishpool, R. M., Southgate, E., Donmoyer, J. E., Walrond, J. P., Moses, J. E. & Kass, I. S. (1978). Structure and physiological activity of the motoneurons of the nematode *Ascaris. Proc Natl Acad Sci USA, 75*(7), 3493–3497.

Takemoto, T. (1978). Isolation and structural identification of naturally occurring excitatory amino acids. In E. G. McGeer, J. W. Olney & P. McGeer (Eds.), *Kainic Acid as a Tool in Neurobiology* (pp. 1–15). New York: Raven Press.

Terada, M., Ishii, A. I., Kino, H. & Sano, M. (1982). Studies on chemotherapy of parasitic helminths (VII). Effects of various cholinergic agents on the motility of *Angiostrongylus cantonensis. Jpn J Pharmacol, 32*(4), 633–642.

Thompson, D. P., Klein, R. D. & Geary, T. G. (1996). Prospects for rational approaches to anthelmintic discovery. *Parasitology, 113*(Suppl 3), S217–238.

Tornoe, C., Holden-Dye, L., Garland, C., Abramson, S. N., Fleming, J. T. & Sattelle, D. B. (1996). Lophotoxin-insensitive nematode nicotinic acetylcholine receptors. *J Exp Biol, 199*(Pt 10), 2161–2168.

Tracy, J. W. & Webster, L. T. (1995). Drugs used in the chemotherapy of Helminthiasis. In J. G. Hardman & L. E. Limbird (Eds.), *Goodman & Gilman's The Pharmacological Basis of Therapeutics* (9th ed., pp. 1009–1026). New York: McGraw Hill.

Unwin, N. (1993). Nicotinic acetylcholine receptor at 9 Å resolution. *J Mol Biol, 229*(4), 1101–1124.

Van den Bossche, H. (1980). Peculiar targets in anthelmintic chemotherapy. *Biochem Pharmacol, 29*(14), 1981–1990.

Van den Bossche, H. & De Nollin, S. (1973). Effects of mebendazole on the absorption of low molecular weight nutrients by *Ascaris suum. Int J Parasitol, 3*(3), 401–407.

Varady, M., Bjorn, H., Craven, J. & Nansen, P. (1997). *In vitro* characterization of lines of *Oesophagostomum dentatum* selected or not selected for resistance to pyrantel, levamisole and ivermectin. *Int J Parasitol, 27*(1), 77–81.

Weiger, W. A. (1997). Serotonergic modulation of behaviour: A phylogenetic overview. *Biol Rev Cambridge Phil Soc, 72*(1), 61–95.

Wood, P. J. & Mansour, T. E. (1986). *Schistosoma mansoni:* Serotonin uptake and its drug inhibition. *Exp Parasitol, 62*(1), 114–119.

Xu, M., Molento, M., Blackhall, W., Ribeiro, P., Beech, R. & Prichard, R. (1998). Ivermectin resistance in nematodes may be caused by alteration of P-glycoprotein homolog. *Mol Biochem Parasitol, 91*(2), 327–335.

8

Targets in the Tegument of Flatworms

Trematodes and cestodes have their external surface covered with an unusual structure termed the tegument. In addition to protecting the parasite from adverse conditions in the host, it has many other functions. These include evasion of the host immune system, the absorption of certain nutrients and the excretion of some metabolic products, control of motility, and control of electrochemical and osmotic gradients. Studies on the structure and biochemistry of the tegument have lately been emphasized when it was discovered that the tegument is the main target for some important antischistosomal agents such as praziquantel and metrifonate.

Most of our information on the tegument in trematodes came from studies on the liver fluke *Fasciola hepatica* (Threadgold, 1963) and the blood fluke *Schistosoma mansoni* (Hockley, 1973; Hockley & McLaren, 1973; McLaren & Hockley, 1977). A recent comprehensive review on the teguments of cestodes and trematodes as well as the cuticle in nematodes is also recommended (Thompson & Geary, 1995). This chapter focuses on trematodes and cestodes.

Structure and Function of Schistosome Tegument

Figure 8.1 is a schematic diagram of the typical components of the dorsal tegument of an adult male *S. mansoni* determined from electron microscopic transmission (Hockley, 1973). The tegument consists of cytoplasmic syncytium 2–4 μm thick, which is a mass of dense granular material that contains no separate cells. The syncytial zone of the tegument is separated from the muscle layer and subtegumental cells by a membranous basal lamina, a regular trilaminate lipid bilayer. The outermost layer of the tegument in *S. mansoni* is heptalaminate consisting of two lipid bilayers. The family *Fasciolidae* have only a single outer lipid bilayer (McLaren & Hockley, 1977). It has been suggested that the additional

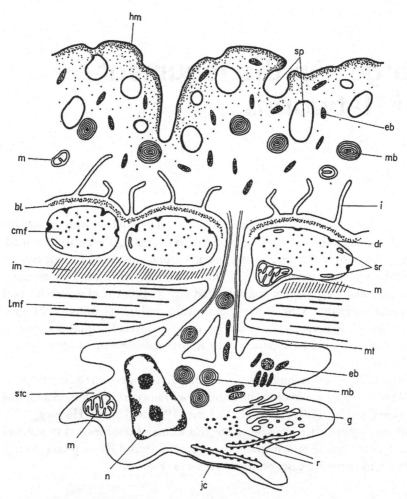

Figure 8.1. Diagram of a typical part of the adult schistosome tegument and its under-lying structures. bl, basal lamina; cmf, circularly arranged muscle fibers; dr, peripheral dense region of muscle fibers; eb, elongate body; g, Golgi; hm, heptalaminate outer membrane; i, invagination of basal membrane; im, interstitial material; jc, junctional complex with parenchymal cell; lmf, longitudinal muscle fiber; m, mitochondrion; mb, membranous body; mt, microtubule; n, nucleus; r, ribosomes; sp surface pits; sr, sar-coplasmic reticulum; stc, subtegumental cell (cyton). Reprinted from Hockley (1973) with permission.

tegumental bilayer is part of the defense mechanism for schistosomes that live in the blood vessels, where they are constantly exposed to the host's immune de-fense system. The exterior heptalaminate membrane starts to develop from the cercarial trilaminate outer membrane about 30 min after cercarial penetration of the host skin. The change from the trilaminate cercarial membrane to the heptalaminate adult membrane is completed in the schistosomulum (the next developmental stage) within 3 hours. The tegumental syncytium contains trilaminate vesicles that contain membranous material and are either spherical or elongated oval in shape. They are referred to as membranous bodies and elongate bodies respectively. The syncytium is connected to nucleated cell bodies (cytons) by cytoplasmic tubes that are lined with microtubules. The cytons are located beneath the circular and longitudinal muscle fibers and are not considered to be part of the tegument. They contain nuclei, mitochondria, ribosomes, Golgi complexes, and glycogen particles. Biogenesis of new membrane material occurs in the cytons. Both membranous bodies and elongate bodies carrying membranous material are found scattered in the cytons in association with Golgi complexes where they are probably formed. These vesicles carry membranous material synthesized within the cytons and migrate through the cytoplasmic tubules to the syncytium and then to the outer membrane surface of the tegument. This is where the contents of the membrane vesicles are incorporated with other tegumental proteins to make a new heptalaminate membrane. This process is particularly needed not only by the developing schistosomulum but also by adult parasites following shedding or damage to the outer surface of the adult.

Schistosome spines are located within the tegumental outer and basal membranes with the pointed tip beyond the general level of the tegument (Matsumoto *et al.*, 1988; see Fig. 8.2). When the schistosome is living in the mesenteric blood vessels of the host the spines are embedded in the endothelial cells of the blood vessels. It has been suggested by Hockley (1973) that these spines hold the worm against the blood flow. The spines are composed of paracrystalline arrays of actin filaments. Actin is also present in regions where tegumental damage and healing occurred in response to immunological, environmental, or chemical stresses. Paramyosin, a cytoskeletal protein not found in the vertebrate host, is found (primarily in a nonfilamentous form) in the cytons and the elongate bodies of the tegument (Matsumoto *et al.*, 1988). Paramyosin has been tested in mice as an immunogen to reduce the infection rate (Lanar *et al.*, 1986). Although paramyosin's role in the schistosome tegument has not yet been identified, it has been suggested to have diversified functions that may include forming a scaffold to which cortical myosin molecules attach. This protein is of special interest as a candidate target for antischistosomal vaccines or chemotherapeutic agents.

The outer membrane of the tegument has many invaginations (termed surface pits), which have the effect of increasing the surface area of the parasite at least tenfold. Such an increase in the surface area provides the parasite with greater opportunity for absorbing nutrients such as glucose and other small molecules from the outside milieu. The outermost membrane bilayer of the schistosomes is shed periodically, with certain domains having a fast rate

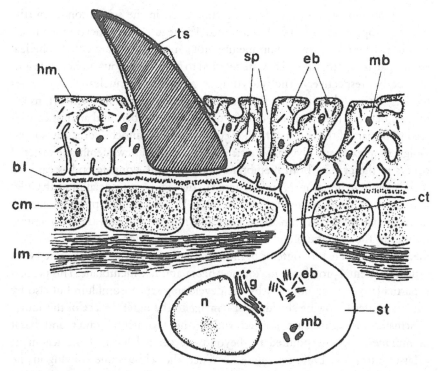

Figure 8.2. Diagram of the dorsal surface of adult male schistosome tegument showing position of the spine(s). bl, basal lamina; cm, circularly arranged muscle fibers; ct, cytoplasmic tube; eb, elongate body; g, Golgi complex; hm, heptalaminate outer membrane; lm, longitudinal muscle fibers; mb, membranous body; n, nucleus of subtegumental cell; sp, surface pits; st, subtegumental cell (cyton); ts, tegumental spine. Reprinted from Matsumoto *et al.* (1988) with permission.

($t_{1/2} = 11$ hours), whereas others shed at an extremely low rate (Bennett & Caulfield, 1991; Caulfield *et al.*, 1991). Shedding of these surface antigens appears to be used by the parasite as a device for immune evasion.

Some Proteins Identified in Schistosome Tegument

Part of the function of the tegument is to allow the parasite to evade the immune system of the host (Abath & Werkhauser, 1996). Schistosomes are continually shedding their tegumental outer surface, particularly those areas that are damaged or that have immunocomplexes formed on the external membrane. Another evasive mechanism is antigenic mimicry. Antigens acquired from the host or synthesized endogenously by the parasite and incorporated on the parasite surface protect the parasite against the immune response of the host. Human monocyte-derived macrophages are lysed by the schistosomula and fail to kill the parasites after their activation by γ-interferon. Both monocyte-derived macrophages and platelets fuse with the outer tegumental membranes of the parasite, providing defensive shields against immune system recognition (Remold *et al.*, 1988).

Glutathione S-transferase (GST) is an enzyme that is essential for the synthesis of reduced glutathione (GSH), which is known to protect schistosomes against host oxidants. GST has been identified in the tegument as well as in the parenchyma, the esophageal epithelium, and the genital organs (Liu *et al.*, 1996). Capron and his group found that immunization with this enzyme gives partial protection against infection from several species of schistosomes (Boulanger *et al.*, 1995).

ATP diphosphorylase, an enzyme that hydrolyzes ATP and ADP to AMP and P_i is also found in the schistosome tegument. ADP is an attractant to blood platelets of the host that are harmful to the parasite. The main function of schistosome ATP diphosphorylase is to evade the effect of ADP accumulation (Vasconcelos *et al.*, 1996). Other proteins that have been identified in the tegument include alkaline phosphatase, β-glucuronidase, amino peptidase, acetylcholinesterase, phosphofructokinase, two glucose transporters, and several glycolytic enzymes. Sensory organelles are located in the tegument and they are thought to be connected to the nervous system and function as mechano receptors (Hockley, 1973).

Two glycolytic enzymes, triose phosphate isomerase (TPI) and glyceraldehyde-3P-dehydrogenase (GAPDH), are tegumental proteins that have been shown to have a partial protective effect against schistosome infection by acting as immunogens (Goudot-Crozel *et al.*, 1989; Harn *et al.*, 1992). Using electron microscopy and immunogold localization in cryosections, phosphofructokinase (PFK) was identified in adult and schistosomular tegument and in adjacent tissues. The enzyme showed high immunogenicity in mice. However, unlike the case with TPI and GAPDH it did not provide significant reduction in worm burden during the course of schistosome infections in mice. One main difference between PFK and the other two enzymes appears to be the inaccessibility of PFK to the outside surface of the tegument (Mansour *et al.*, 2000). The role played by these enzymes in glycolysis in the tegument has not been established.

Protein Attachment to the Tegument

The exterior surface of the tegument of schistosomes and schistosomula contain many enzymes and other proteins that can be released by other enzyme

systems. These include acetylcholinesterase, alkaline phosphatase, and other serine hydrolases that have been identified on the exterior surface of the tegument
of schistosomules as ectoenzymes (Espinoza *et al.*, 1995). The attachment of
these surface proteins to the tegument is through a hydrophobic anchor, the
1,2-diacylglycerol moiety of a phosphatidylinositol (PI) molecule covalently attached to one end of the protein via an intervening oligosaccharide. The surface
proteins can be released from the live schistosomular tegumental membrane by
a phosphatidylinositol phospholipase C (PIPLC). Levels of the released tegumental enzymes reach a plateau in schistosomula that are 24-hours old, a time
when the outside tegumental membrane is heptalaminate and fully developed.
The identification of acetylcholinesterase and other serine hydroxylases in the
outer membrane of the tegument and their release by PIPLC are reminiscent of
the variant surface glycoproteins (VSGs) of *T. brucei*, which have been shown to
be associated with similar membrane-anchoring domains (Ferguson, Haldar,
& Cross, 1985). Glycosylphosphatidylinositol specific lipase activities from
Fasciola hepatica and *Schistosoma mansoni* were identified in detergent extracts of
both species of adult worms as well as in the cercariae of schistosomes (Hawn &
Strand, 1993). The relationship of proteins that are anchored outside the schistosome membranes to the control of glucose transport and to the mechanisms
by which the parasites evade the immune system are the subject of many investigations. Some of these tegumental proteins function as glucose transporters
across the tegument.

Glucose Transport across the Tegument

The importance of the role that glucose metabolism plays in the survival of
S. mansoni, F. hepatica, and related parasites has been discussed in Chapter 3. To
obtain sufficient energy for survival in the host these parasites have to utilize
massive amounts of glucose, which is taken up primarily through the parasite
tegument and not by the intestinal caecum. The tegument is also important in
excretion of metabolic products (Mansour, 1959; Rogers & Bueding, 1975). The
trematode tegument is the primary absorptive surface for other low molecular
weight solutes including amino acids and cholesterol as well as glucose (Pappas,
1988). The cestodes have no intestine so they depend entirely on absorption
through the tegument for all their metabolic needs.

Glucose transport across the tegument is facilitated by glucose transporters. Shoemaker and his group have identified these transporters in schistosome tegument. Two cDNA clones were isolated that encode for two different

glucose transporters in the tegument; these were designated schistosome glucose transporter 1 (SGTP1) and 4 (SGTP4) (Skelly *et al.*, 1994). There is 60% similarity between their amino acid sequences. The two transporters were also identified by electron microscopy and found to have different locations in the tegument: SGTP1 is present at the basal lamina (see Fig. 8.1) of the tegument and to a lesser extent in the underlying muscle cells (Zhong *et al.*, 1995). The location of SGTP1 at that site indicates that its main function is to transport glucose from the tegumental matrix through the extracellular space to the cells directly beneath it. Whereas SGTP1 transporters are part of a typical bilayer membrane, the SGTP4 is located at the outer membrane of the tegument, which has an exceptional double lipid bilayer structure.

The distinct location of SGTP4 at the outer tegumental membrane indicates that these are the transporters that facilitate the transfer of glucose from the glucose-rich host blood into the parasite tegument. SGTP4 transporters are evenly distributed throughout both the ventral and dorsal surfaces of males and females (Skelly & Shoemaker, 1996). SGTP4 transporters are developmentally controlled and they appear during the transformation of the free-living cercariae into schistosomula during maturation. Their appearance coincides with the time that the schistosomula change from a trilaminate outer membrane to a heptalaminate double lipid bilayer membrane just after they penetrate through the host skin. The SGTP4 appears to satisfy the needs of the parasites for high glucose uptake as soon as they enter the host (schistosomula stage) and throughout adulthood.

The different stages of the biosynthesis of SGTP4 appear to follow the same route described for other outer membrane structures of the tegument. Using immunofluorescence microscopy Skelly and Shoemaker (2001) showed that SGTP4 is detected within minutes of initiation of transformation of cercariae into schistosomula and is localized to a cyton network beneath the musculature. Thirty minutes after the initiation of transformation the transporters are seen in tubules connecting the cytons to the tegumental surface. Within 3 hours SGTP4 was shown to be largely covering most of the developing worm surface. The intermediate steps, from the cytons to outer membranes, which include the appearance of SGTP4 within discoid bodies and multilamellar vesicles (referred to in Fig. 8.1 as membranous bodies and elongate bodies), were identified using electron microscopy (Jiang et al., 1996). SGTP4 serves a vital function in the life of schistosomes. The final location of these sugar transporters at a distinct site at the surface of the tegument makes them a good target for the development of antiparasitic agents or a protective vaccine.

Tegumental Acetylcholine Receptors, Acetylcholinesterase, and Glucose Uptake by Schistosomes

The participation of acetylcholine in regulating motility and synaptic transmission in parasitic flatworms has been discussed in Chapter 7. Recently, a new site of acetylcholine function at the schistosome tegument has been under investigation by several laboratories. Both acetylcholinesterase (AChE) and a nicotinic type of acetylcholine receptor (nAChR) were identified at the outer side of the tegumental membrane. These proteins are predominantly concentrated on the dorsal surface of the adult male. Neither enzyme nor receptor are prevalent on the head, tail, or ventral surface of the adult male. The adult female, which normally lodges in the gynaecophoral canal of the male, has only a small amount of these proteins (Camacho *et al.*, 1995).

Evidence for the involvement of AChE and nAChR in the process of glucose uptake was reported (Camacho & Agnew, 1995b). Exposure of *S. hematobium* or *S. bovis*, but not *S. mansoni*, to the low concentrations of acetylcholine (10^{-8} to 10^{-9} M) that could be found in the host blood enhances glucose uptake by the parasites. However, at higher concentrations (10^{-5} to 10^{-6} M) acetylcholine inhibits glucose uptake by the parasites. The specificity of the interaction between acetylcholine and the nicotinic receptor was strengthened by showing that the effect of acetylcholine was antagonized by specific antagonists such as d-tubocurarine and α-bungarotoxin. The fact that α-bungarotoxin antagonizes the effect of acetylcholine indicates that the nicotinic receptor is on the tegument because this inhibitor does not penetrate the tegument. Further evidence for a relationship between the cholinergic receptors and glucose uptake came from studies on the antischistosomal effect of metrifonate, a known inhibitor of AChE.

Monoclonal antibodies against AChE of *S. mansoni* have been prepared and evidence of differences between parasite and host acetylcholinesterase has been reported. Some of the monoclonal antibodies interact only with parasite acetylcholinesterase, but not with the vertebrate enzyme, suggesting that the enzymes have different epitopes. Also, monoclonal antibodies interact with intact schistosomula and cause complement-dependent cytotoxicity (Espinoza *et al.*, 1995). Both tegumental and muscle acetylcholinesterase from *S. hematobium* have been studied (Camacho, Alsford, & Agnew, 1996). The enzyme is present in the parasites in only one form, the dimeric globular form (G_2). Larger forms of the enzyme have been identified, but these appear to be aggregates of the dimeric globular forms. This is unlike the acetylcholinesterase of the host, which occurs in globular monomers, dimers, and tetramers. In the parasites the major part of the AChE is membrane associated. Release of the AChE from tegument by

PIPLC is coupled to a compensatory *de novo* synthesis of AChE to replenish the enzyme that is removed from the surface of the parasites.

The findings of Agnew and her associates of a connection between glucose uptake by the schistosomes and their tegumental AChE/nAChR uncovers a new control mechanism for acetylcholine that is not related to neuromuscular function. There are recent reports of acetylcholine effects on various other cellular functions that are not neuromuscular in nature. For example, it has been demonstrated that acetylcholine and related choline esters acting on a nicotinic acetylcholine receptor can activate cation conductance channels that are responsible for long-term maintenance of low red-cell-membrane potential (Bennekou, 1993). Furthermore, a Na^+/glucose cotransporter from mammalian small intestine appears to function as a negatively charged carrier. A negative membrane potential facilitates sugar uptake by the transporters (Parent *et al.*, 1992). Molecular studies indicate that these cotransporters are present in mammals, fish, and marine mussels. There appears to be evolutionary conservation of both structure and function of these Na^+/glucose cotransporters (Pajor, Hirayama, & Wright, 1992). There is also an indication of involvement of acetylcholine receptors in the process of cell proliferation in human lymphocytes (Richman & Arnason, 1979). Carbamylcholine at micromolar concentrations increases cell proliferation. This cholinergic effect is blocked by atropine and is therefore considered pharmacologically muscarinic. Therefore, both nicotinic acetylcholine receptors and muscarinic have been shown to be involved in at least three cellular regulatory processes that are not part of neuromuscular synaptic transmission.

Metrifonate (Bilarcil) as AChE Inhibitor and Antischistosomal Agent

Metrifonate is an organophosphate inhibitor of cholinesterases. It is used as an alternative drug for the treatment of *S. hematobium* if the infection does not

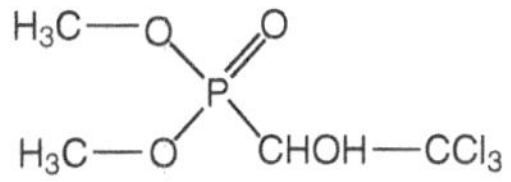

METRIFONATE

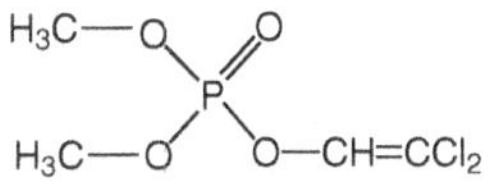

DICHLORVOS

respond to praziquantel. It is not effective against *S. mansoni* infections. It is a prodrug as the anthelmintic activity is due to spontaneous rearrangement of the compound to form dichlorvos (2,2,-dichlorovinyl dimethyl phosphate); which is a better inhibitor of schistosome cholinesterase. The antiparasitic effect of metrifonate, therefore, appears to be due to dichlorvos rather than to metrifonate itself. See the review by Marshall (1987). There is now evidence to indicate that the target of metrifonate and its derivative dichlorvos is AChE in the tegument of the schistosome. This is based on the discovery that acetylcholine at higher concentrations inhibits glucose uptake by the parasite (Camacho & Agnew, 1995a). It therefore appears that under ordinary physiological conditions the low levels of ACh in the blood should favor an increase in glucose uptake by the parasites. *In vitro* experiments carried out on intact *S. hematobium* and *S. mansoni* showed that both metrifonate and dichlorvos inhibit AChE from either parasite (Bueding, Liu, & Rogers, 1972). Metrifonate and dichlorvos presumably increase the level of acetylcholine to concentrations that are inhibitory to glucose transport through the tegument. Such inhibition deprives the schistosome of its prime source of energy and should lead to a lethal effect on the parasite.

The known clinical efficacy of metrifonate against *S. hematobium*, but not against *S. mansoni*, appears to be related to other factors in addition to inhibition of acetylcholinesterase. In single-species schistosome infections *S. mansoni* eggs are usually found in the stool, whereas in *S. hematobium* infections the eggs are excreted in the urine. The effect of metrifonate on humans with mixed infections of *S. mansoni* and *S. hematobium* has been studied (Doehring, Poggensee, & Feldmeier, 1986). In these human cases the eggs from both species appear in the urine and the stool, indicating that both species are inhabiting the perivesical venules of the urinary bladder and the mesenteric venules of the intestine. Treatment of these patients with metrifonate resulted in lowered egg count from both species in the urine. However, egg excretion from either species through the stool was not influenced by the drug. These results indicate that there are other factors, such as the location of the parasites in the host, that modify the effects of metrifonate.

Potential Studies on Tegumental Nicotinic Targets

An argument can be made that the function of the acetylcholine receptors in the tegument must be of a different nature when compared to those in the schistosome cholinergic neurosynapses. The findings by Agnew that ACh increases the rate of glucose uptake of schistosomes indicate that specific cholinergic receptors in the tegument may be involved in this process. Because the parasites are dependent for their survival on glucose metabolism, further information about this system deserves a great deal of attention. This parasite acetylcholine receptor has been identified by fluorescent techniques and selective binding of α-bungarotoxin to be restricted to the tegument on the schistosome surface (Camacho *et al.*, 1995). The receptor should be cloned and expressed in *Xenopus* oocytes to better assess its electrophysiological properties. There are large numbers of both cholinergic agonists and antagonists available in commercial and academic laboratories. Cloned schistosome AChR could be used for screening prospective selective modifiers. Acetylcholinesterase is another component of the cholinergic system that deserves more investigation, and its role in the tegument functions of the parasite should be determined. The enzyme was shown in Arnon's laboratory to be a component of the parasite surface and is anchored to the membrane via covalently attached phosphatidylinositol (Espinoza *et al.*, 1988). Schistosome phosphatidylinositol-specific phospholipase C can endogenously release the AChE from the parasite tegument. This enzyme also deserves further study of its properties and the effect of other inhibitors on its activity. Clearly, metrifonate is not the ideal selective drug that acts on this system.

Praziquantel (PZQ)

Praziquantel is the most important addition to the group of antischistosomal agents. For a review see King & Mahmoud (1989). PZQ is a tetracyclic pyrazino-isoquinoline derivative. The anthelmintic activity is conferred by the pyrazino[2,1-a] isoquinoline moiety. It is highly effective against the three main schistosome species in humans, *S. mansoni*, *S. hematobium*, and *S. japonicum*, as well as other minor species, *S. mekongi*, and *S. intercalatum*. In addition to its superior efficacy against schistosomes, it can be given by mouth as a single dose. This is an important advantage over many other antischistosomal agents that have to be given parenterally in repeated doses. It is therefore valuable and convenient for the therapy of large populations of patients in developing countries. Praziquantel is also effective against intestinal fluke infections and infections of *Clonorchis* and *Fasciola* (liver flukes), paragoniamiasis (lung flukes), and cysticercosis (tape worm larvae). In addition, infections with different species of cestodes (tapeworms) respond favorably to treatment with PZQ. The drug has also been shown to be effective against many trematodes and cestodes that infect domestic animals (Campbell & Garcia, 1986). Prazinquantel is therefore considered to be a broad-spectrum anthelmintic.

Mechanism of Action of Praziquantel

Early observations on the mechanism of action of praziquantel on schistosomes demonstrated an *in vitro* effect on motility of the parasite characterized by rapid tonic contraction of the organism followed by flaccid paralysis. These effects were attributed to the interaction of praziquantel with specific Ca^{2+}-permeable sites in the tegument and muscle cells of the parasite. An increase in the influx of calcium follows and is probably the cause of this effect on motility (Blair, Bennett, & Pax, 1992).

The *in vivo* effects of praziquantel on the tegument were studied by scanning electron microscopy (Shaw & Erasmus, 1983). Administration of low curative doses of praziquantel shows different degrees of surface damage. The damage is characterized by the formation of surface blebbing, with some swelling, vesication, and vacuolization. Eventually there is disintegration of the tegument surface and migration of host leucocytes to these lesions (Mehlhorn *et al.*, 1981). The tegumental lesions are particularly noticed at the surface spines (Fig. 8.2) and the tubercles (papillae present in the dorsal tegument) of the male schistosomes. The intensity of tegumental damage caused by praziquantel depends on the dosage and the length of time after treatment.

The effects of praziquantel indicated above include the brief stimulation of motility and subsequent paralysis and its damage to the tegument, but there is no definitive information on the primary binding sites of the drug or on how it produces this serious damage. Subsequent research on the mechanism of action of praziquantel showed that its efficacy against schistosome infections in mice depends on participation of the host immune system. Schistosomes and related helminths live in the mammalian host circulatory system for many years (reportedly up to twenty years) practically unaffected by the host's immune system (Harris, Russell, & Charters, 1984). The parasite tegument provides the first line of defense for evading attack by the host's immune system. In the absence of any drug treatment, the schistosome outer surface acquires glycolipids, immunoglobulins, and erythrocyte proteins, which bind to the outer surface of the tegument masking parasite antigens and making them inaccessible to the immune responses of the host. Damage induced by PZQ to the tegument appears to interfere with the parasite's evasion of its host's immune responses. Antigens on the schistosome surface that are normally hidden from the host's immune system are exposed and can be recognized by the host.

The synergistic effect between PZQ and the immune system was shown in experiments using immuno-compromised mice. For a review see Brindley & Sher (1990). PZQ was less effective as an antischistosomal agent in adult thymectomized mice. In addition, antibodies were shown to be necessary mediators in the synergy between the drug and the immune response. The efficacy of PZQ was also reduced in schistosome-infected mice that were B-lymphocyte-depleted (μ-suppressed). In these experiments the antiparasitic effect of praziquantel was completely restored in the μ-suppressed mice by passive transfer of immune serum from donor mice that were infected with schistosomes. These experiments provide further evidence that the host's antibody response is essential for the antiparasitic effects of praziquantel (Brindley & Sher, 1987).

Antigens That Attract the Host's Immune System

What is the nature of these tegumental antigens that attract host antibodies? Nude (athymic) mice were used as hosts for schistosomes (Brindley et al., 1989). These mice do not produce antibodies that would interfere and mask identification of the immunofluorescent antibodies that are being produced in response to praziquantel. A panel of monoclonal antibodies raised against several tegumental and subtegumental proteins were tested for their reactivity to both treated and untreated worms. Antibodies do not normally bind to the surface of intact untreated parasites. It was found, however, that five of the antitegumental antibodies tested bound *in vitro* to the surface of parasites removed from nude mice 1 hour after PZQ treatment. Presumably these antibodies are being produced to parasite tegumental proteins that were exposed in response to praziquantel. One molecule of special interest was a 200-kDa glycoprotein that is present in high concentrations in the tubercles and parenchyma of adult *S. mansoni*. These proteins gave an immunofluorescent pattern that is similar to those observed in schistosomes removed from praziquantel-treated infected mice. Further evidence to implicate this protein as a major target of the immune response was established by showing that passive transfer of one of the 200-kDa positive monoclonal antibodies to immunocompromised mice reconstitutes the efficacy of praziquantel as an anthelmintic. Biochemical characterization of the 200-kDa glycoprotein shows that it is linked to the membrane of the adult schistosome by a glycosylphosphatidylinositol anchor (Sauma & Strand, 1990). Only part of the 200-kDa glycoprotein is exposed to the immune response during drug treatment. The 200-kDa antigen has been cloned and the deduced amino acid sequence does not match any of 400 known membrane protein motifs. The recombinant protein encoded by the cDNA clones expressed epitopes that are similar to those expressed on the surface of PZQ-treated worms. This was verified by showing that antibodies to fusion proteins encoded by cDNA could recognize epitopes exposed on the surface of worms by PZQ treatment (Hall, Joseph, & Strand, 1995). Praziquantel exposes epitopes of this peptide on the worm surface that are normally not exposed.

Another antigen that has been identified as being exposed after PZQ treatment is a 27-kDa protein. It is also abundant in the tubercles of adult male schistosomes and possesses nonspecific esterolytic activity (Doenhoff, Modha, & Lambertucci, 1988). This protein has been proven to be involved in the effect of PZQ against schistosome infection. Rabbit antiserum raised against the protein was found to augment the efficacy of PZQ against parasite infection in mice.

A recent study on the PZQ-induced surface lesions was reported (Linder & Thors, 1992). Actin, a major muscle protein, was found to be exposed at the surface of the worm following *in vivo* or *in vitro* exposure to PZQ. No exposed actin could be detected on adult worms from untreated mice. Actin depolymerization, leading to loss of dorsal surface integrity, was observed. Schistosomes isolated from PZQ-treated animals were shown to have large amounts of bound Ig in association with their tegumental tubercles. Disintegration of the tegumental spines (Fig. 8.2) *in vivo* eventually occurs and was shown to be associated with binding of host antibodies against actin. This effect appears to be linked to binding of an actin depolymerizing protein (gelsolin) from the host. Gelsolin is a 20-kDa calcium-regulated actin binding protein found in the plasma that controls the depolymerization of F-actin. Gelsolin severs the actin filament and caps its growing end. The distribution sites of bound gelsolin coincide with those of actin. By using immunofluorescent staining with antigelsolin antibodies gelsolin in human plasma was shown to bind to frozen sections of the parasites. Two processes appear to be involved in the loss of surface spines following PZQ treatment. The first is the binding of antibodies against actin directed against the schistosome surface membranes damaged by the drug; the second involves the action of the host's actin depolymerizing protein, gelsolin.

Although no adequate research has been done on the effect of PZQ on other trematodes nor against cestodes, the evidence from a few cases suggests that synergy between the anthelmintic and the immune response of the host is necessary for its lethal effect. The similarity of the structure and biochemistry of teguments among different trematodes and cestodes supports this view (Thompson & Geary, 1995).

Resistance of Schistosomes to PZQ

Soon after PZQ was introduced as the drug of choice for the treatment of all types of human schistosomiasis the question of whether the parasites can become resistant to its effect became controversial. Several reports from different parts of the world where schistosomiasis is endemic claimed or denied the emergence of resistant strains (WHO, 1992). Therapeutic failures after extended treatment of human infections were reported, but these had been attributed to rapid reinfection, presence of immature parasites at the time of drug treatment, and low levels of the antischistosome antibodies required for high efficacy of PZQ. Experiments using mice to examine the susceptibility of *S. mansoni* to PZQ were carried out (Fallon & Doenhoff, 1994). Mice were treated with multiple subcurative doses of PZQ and eggs of surviving worms were used for life-cycle passage. Six or seven drug-treated passages of the parasites in mice were carried out using increasing doses of PZQ with each passage. Eighty percent of the worms that survived the sixth passage of selection for PZQ resistance survived three doses of 300 mg/kg of PZQ. This is considered to be high resistance to the drug. In contrast, in the control group that had not been subjected to any drug pressure, survival of worms was only 11% with the same dose of PZQ. These experiments indicated for the first time that drug resistance to PZQ can be transmitted to a parasite population that was previously sensitive to the antischistosomal agent. Thus, schistosomes in mice have the potential to become resistant to PZQ and the resistance can be carried over to the next life cycle of the parasite as it is transmitted to the mammalian host.

The antiparasitic effect of PZQ is a synergy between the host's immune response and the direct effect of the drug on schistosome tegument and musculature. The question arises whether demonstrated resistance is primarily due to host factors that involve the response of the immune system or due to a direct effect on the parasites. Recent experiments were carried out on isolates from patients that did not respond to treatment with three doses of PZQ. These isolates were used to produce experimental infections in mice. Eight out of twelve isolates showed resistance to PZQ in mice when compared to the control group. The effect of PZQ on contractility of PZQ-resistant male schistosomes was examined in an apparatus that measures muscle tension. *In vitro* contractile responses to PZQ of parasites from resistant isolates were found to be significantly reduced when compared to control parasites. Thus isolates that produce resistant infections in mice show diminished contraction in response to PZQ *in vitro*. This study directly correlates resistant isolates of schistosomes with contractility without participation of mammalian hosts (Ismail *et al.*, 1999).

In summary, the clinical reports from different geographical areas where PZQ has been used indicate that human schistosomiasis patients are exhibiting signs of resistance to its effectiveness as the prime antischistosomal agent. Moreover, several laboratory reports indicate that the parasite itself, without host participation, shows signs of resistance to PZQ. Since the drug is now becoming more available to worldwide programs to eradicate the disease it is anticipated that selection for drug-resistant schistosomes will continue on a larger scale. In many countries (e.g., Egypt) there is already great reliance on PZQ as the only drug against schistosomiasis. This new information makes one wonder about the future effectiveness of PZQ for the treatment of the 200 million schistosomiasis patients worldwide. Rather than relying mostly on clinical reports we need to establish a system to carefully monitor the resistance of the parasites themselves to PZQ, based on testing different isolates in mice. New antischistosomal agents that are as effective and as safe as PZQ are badly needed, for the time will come when PZQ loses its effectiveness as the prime drug choice against schistosomiasis.

Potential Studies on Ca^{2+} Influx across the Tegument

Recent research on the mechanism of action of PZQ has increased interest in the tegument of schistosomes. There is now general consensus that the effect of PZQ involves an increase in Ca^{2+} influx across the tegument, resulting in muscle contraction followed by tegumental damage and attraction of the host's inflammatory cells to the parasite (Redman et al., 1996). In spite of these tegumental changes, identification of PZQ receptors remains elusive. Synergy between the host's immune responses and the antischistosomal effect is not a unique feature of PZQ. For example, mice that are T-cell deprived and infected with schistosomes are less sensitive to treatment with other schistosomicidal agents such as potassium antimony tartrate, hycanthone, and oxamniquine, as well as praziquantel, when compared to immunologically intact controls (Sabah et al., 1985).

The question remains as to what triggers the two main effects of PZQ, induction of Ca^{2+} influx that is manifested by muscle contraction and exposure of tegumental antigens that attract host immune cells. The evidence available indicates that the prime target for PZQ action lies in the tegument, but it has not as yet been specifically identified (Fetterer, Pax, & Bennett, 1980; Thompson, Pax, & Bennett, 1982; Wolde Mussie et al., 1982; Redman et al., 1996). Studies on electrical potential in the tegument of adult male schistosomes and the role of Ca^{2+} in muscular contraction revealed that the tegument may be electrically coupled to the muscle layers that lie beneath the tegument. A rise in intrategumental Ca^{2+} could lead to increased Ca^{2+} in muscle cells followed by muscle contraction. Ca^{2+} is a necessary cation for actin/myosin contraction. Spontaneous contractions of the parasite in vitro depends on the presence of calcium. Schistosomes incubated in media without Ca^{2+} have their PZQ-induced contractions reduced. As the endogenous calcium is depleted from the parasite contraction is reduced gradually. The role of calcium on contractility is not restricted to the effect of PZQ. Contraction induced in the parasite by high K^+ or dinitrophenol is gradually reduced following calcium deprivation and contractility resumes when Ca^{2+} is added back to the medium. The role played by calcium in F. hepatica contraction is the same as in schistosomes. Spontaneous contraction in this parasite depends on extracellular Ca^{2+}. Contraction is abolished in Ca^{2+}-free saline and when Ca^{2+} channel blockers such as nifedipine or cadmium chloride are added to the media (Graham, McGeown, & Fairweather,

1999). These findings, when added to other reports, indicate that the primary determinant in contraction of flatworms is extracellular calcium. The details of how Ca^{2+} influx across the tegument is controlled have still to be elucidated.

Several Ca^{2+} channels have been found in the membranes of the tegument and muscle cells that may participate in the entrance of the cation across the outer tegumental membrane (Redman *et al.*, 1996). These include second messenger-gated Ca^{2+} channels, calcium release-activated Ca^{2+} channels, agonist-gated Ca^{2+} channels, and voltage-gated Ca^{2+} channels. None of these channels have been identified experimentally and their responsiveness to PZQ was not determined. A putative Ca^{2+}-transport ATPase gene has been cloned from *S. mansoni* (SMA1; de Mendonca *et al.*, 1995). The membrane topology for SMA1 is characteristic of the P-type ATPases (located in membranes of eukaryotes) and its molecular structure confirms its function as a Ca^{2+} transport ATPase. Future studies on the expressed protein and its responsiveness as a target of PZQ should contribute to a better understanding of Ca^{2+} transport through the parasite's outside bilayer. The expectation is that genes for other transporters will be cloned and their role in Ca^{2+} transport, and their possible participation as targets for PZQ action will be investigated.

Oxamniquine

Oxamniquine is currently used as a second choice against *S. mansoni*, but it has a lesser therapeutic effect against *S. hematobium* or *S. japonicum*. Because of its limited usefulness the antischistosomal effects of oxamniquine have not been investigated as extensively as those of praziquantel. Several mechanisms of action have been suggested, but the current information is that oxamniquine produces damage to the tegument similar to that observed with praziquantel. However, it takes a longer time for the damage to become apparent. Mice infected with *S. mansoni* and treated with oxamniquine were compared with those treated with PZQ. Scanning electron microscopy studies on schistosomes from mice treated with PZQ showed the characteristic disruption of the male schistosome tegumental surface, with blebbing on the dorsal tubercles and protuberances on the worm surface 1 hour after treatment. The greatest degree of disruption of the

OXAMNIQUINE

tegument occurred on the second day of the treatment. However, oxamniquine treatment did not show the same effects until day 4 post-treatment and the teguments were extremely damaged only after 12 days post-treatment (Fallon, Fookes, & Wharton, 1996). The considerably slower appearance of drug damage to the tegument by oxamniquine was reflected in the slower exposure of parasite antigens and also by the lesser adherence of host antibodies to the surface of drug-treated worms. Treatment with either drug demonstrated no damage to the female tegument, but the female schistosomes appeared stunted. Experiments also showed synergy between oxamniquine and serum from rabbits immunized with tegument extract when the extracts were given 6–9 days after drug treatment. This synergistic effect between drug administration and antibody can occur only when sufficient time has elapsed for oxamniquine to damage the tegument. The antischistosomal effect of oxamniquine is presumably the result of tegumental damage followed by attraction of host granulocytes and macrophages to the damaged worm surface.

Dyneins in *Schistosoma mansoni*

There has been great interest in identifying different tegumental proteins and their relationship to parasite survival in the host. Strand's laboratory has studied a series of proteins that appear to function in the maintenance and/or damage repair of the schistosome tegument (Hoffmann & Strand, 1996, 1997). Using monoclonal antibodies generated against antigens associated with tegumental membrane proteins they were able to identify two polypeptides, one of 8.9 kDa and a second of 7.6 kDa. These polypeptides appear to be related to dynein light chain proteins in the flagella of *Chlamydomonas* and in *C. elegans* (King & Patel-King, 1995). Dynein light chains could be part of large enzyme complex assemblies of heavy, intermediate, and light dynein chains. Dyneins are macromolecular motors that attach to the appropriate molecular cargo such as membranous organelles along microtubules (King & Patel-King, 1995). The dyneins are also involved in spindle assembly used for chromosome movement in mitosis. An interesting feature of these proteins is that their expression in *S. mansoni* is developmentally regulated. They have been localized in the schistosomula, after skin penetration and at the lung stage, and in adults. These are the stages that also show early expression of the heptalaminate outer membranes. The dynein proteins are not expressed in the cercariae nor in the ciliated miracidia. The newly discovered dynein light chain proteins have affinity to other tegumental proteins with which they make highly complex assemblies. Another dynein light chain protein has also been characterized as a 20.8-kDa tegumental antigen

(Hoffmann & Strand, 1997). This dynein is also developmentally regulated with the highest concentration being found in cercariae. This polypeptide interacts specifically with a *S. mansoni* 10.4-kDa dynein light chain found previously (Hoffmann & Strand, 1996). Velocity sedimentation analysis indicates that the 20.8-kDa and the 10.4-kDa polypeptides are naturally associated as a complex. This protein appears to be a prospective target either as a vaccine candidate or as a target for chemotherapeutic agents. Although the protein is currently looked at as a vaccine candidate against schistosomiasis (Webster *et al.*, 1996) there has not been enough research on its associated structural proteins and their function in the process of biosynthesis and repair of the heptalaminate membrane. More information is needed about the regulation of members of the dynein family of proteins that appears to be of importance in schistosome survival in the host.

The Effect of Triclabendazole on *Fasciola hepatica* Tegument

Fascioliasis is primarily an infection of sheep and cattle and it was only recently that reports of human infections became more prevalent. It is becoming increasingly important in public health in Egypt, Bolivia, and Ecuador. WHO reports that there are 2.4 million people infected and that 180 million are at risk (WHO, 1997). Recently, there has been an upsurge of cases of human fascioliasis in Egyptian children. PZQ, given in multiple doses, was claimed to be effective, but there have also been failures in using it as a complete cure. Bithionol has been used for some time as a drug of choice but is also not fully satisfactory. Triclabendazole, a benzimidazole, was reported to be effective and well tolerated in human fascioliasis (el-Karaksy *et al.*, 1999).

TRICLABENDAZOLE

Since its success as an anthelmintic against *Fasciola* attempts were made to identify the mechanism of action of triclabendazole (TCBZ). Several members of the benzimidazole group are known to owe their antinematodal effect to a selective depolymerization of cytoplasmic microtubules in cells that are involved in different vital functions of the parasite. *In vivo* tests of the effect of TCBZ on *Fasciola* infections showed that the drug is an effective fasciolicide against both juvenile and adult stages of the parasite. In an attempt to identify the site of action of TCBZ in intact parasites both juvenile (3 weeks post-infection) and adult parasites were incubated with media containing the active sulfoxide metabolite of triclabendazole at concentrations that correspond to blood levels (15 μg/ml). Transmission electron microscopy was used to identify changes in the tegumental surface of the parasite (Stitt & Fairweather, 1994). Signs of severe damage to the parasite tegument were shown at concentrations of 15 μg/ml in both adult and juvenile worms. There was vacuolization, blebbing, syncytium disruption, and convolution of the outer plasma membrane of the tegument. Secretory bodies in the tegument were shown to accumulate in the syncytium. These secretory bodies normally carry glycoproteins to the outer membrane for tegument repair after damage. Many of these signs of the damaged tegument in *Fasciola* are reminiscent of those changes described above in schistosomes treated with praziquantel. Eventually, those changes lead to disruption and sloughing off of the tegument. The ultimate result is loss of tegumental function in nutrient absorption and excretion of metabolic products as well as loss of the parasites' ability to evade the host's immune system. Because microtubules have been visualized immunocytochemically in the tegument the possibility has been raised that there is a link between TCBZ's inhibitory effect on microtubule-based function and the damage to tegument structure (Stitt & Fairweather, 1994); also see Fig. 8.1. An effect of triclabendazole on these microtubules similar to their reported effect on nematodes may be implicated.

Drugs Acting on the Tegument of Cestodes (Tapeworms)

The ultrastructure of the cestode tegument is similar to that described for schistosomes. There is the syncytial layer with its outer membranes, basal lamina, and tegumental cytons. Because cestodes have no intestinal cecum for digestion and absorption of macromolecular nutrients from the host's intestine (e.g., digested peptides) the outer surface of the tegument has structures similar to the "brush border" of the mammalian intestine, which increase the cestode surface 26–30 times. These structures are called microtriche. The question of how large molecular nutrients enter through the tegument or whether cestodes use endocytosis has not as yet been resolved (Threadgold, 1984).

Tapeworms live during their adult stages in the mammalian host's gut. Larvae of some of these parasites can use humans as intermediate hosts. The most clinically prevalent and cosmopolitan of tapeworms include *Taenia saginata* (beef tapeworm), *Taenia solium* (pork tapeworm), and *Hymenolepis nana*. Both *T. solium* and *H. nana* can use the human as intermediate as well as definitive host. Praziquantel is the drug of choice against all tapeworms with niclosamide as second choice. Most of the research on drug effects is done on *Mesocestoides corti* because the larvae of these parasites can easily be maintained in mice by intraperitoneal injection. The life cycle of this parasite is unique in that the adult parasites in the gut can reproduce both sexually and asexually.

Praziquantel, when applied to cultures of *Mesocestoides corti*, elicits an increase in contractions of the parasites and subsequent morphological damage to the surface of the parasites. This is characterized by erosion of the surface microvillous layer. When the drug is given in a subcurative dose to *M. corti*–infected mice there is a significant reduction in numbers of the parasites and an increase in phagocytic activity of peritoneal macrophages. The reduction in parasite number correlates well with the increase in phagocytic activity. Just as in the case of schistosomes and PZQ, there is a synergy between drug damage to the tegument and immune responses of the host (Hrckova & Velebny, 1997; Hrckova, Velebny, & Corba, 1998).

The effect of niclosamide on the tapeworm tegument appears to be similar to that of praziquantel. The drug, when given to mice infected with *Hymenolepis fraterna*, induces pathomorphological changes in the tegument such as blebbing and eventual erosion of the brush border. Mature and gravid posterior parts of the parasite are more sensitive to the drug damage than those in the anterior strobila and neck (Stoitsova, Gorchilova, & Danek, 1992).

The prime role played by the tegument in the life of platyhelminths has prompted many investigators to study tegument structure and function. Enzymes and structural proteins involved in the function of this shield are very attractive targets for chemotherapeutic agents. Studies on the teguments of other platyhelminths would also contribute to finding better cures for these diseases.

REFERENCES

Abath, F. G. & Werkhauser, R. C. (1996). The tegument of *Schistosoma mansoni*: Functional and immunological features. *Parasite Immunol, 18*(1), 15–20.

Bennekou, P. (1993). The voltage-gated non-selective cation channel from human red cells is sensitive to acetylcholine. *Biochim Biophys Acta, 1147*(1), 165–167.

Bennett, M. W. & Caulfield, J. P. (1991). Specific binding of human low-density lipoprotein to the surface of schistosomula of *Schistosoma mansoni* and ingestion by the parasite. *Am J Pathol, 138*(5), 1173–1182.

Blair, K. L., Bennett, J. L. & Pax, R. A. (1992). Praziquantel: Physiological evidence for its site(s) of action in magnesium-paralysed *Schistosoma mansoni*. *Parasitology, 104*(Pt 1), 59–66.

Boulanger, D., Warter, A., Trottein, F., Mauny, F., Bremond, P., Audibert, F., Couret, D., Kadri, S., Godin, C., Sellin, E. et al. (1995). Vaccination of patas monkeys experimentally infected with Schistosoma haematobium using a recombinant glutathione S-transferase cloned from S. mansoni. Parasite Immunol, 17(7), 361–369.

Brindley, P. J. & Sher, A. (1987). The chemotherapeutic effect of praziquantel against Schistosoma mansoni is dependent on host antibody response. J Immunol, 139(1), 215–220.

Brindley, P. J. & Sher, A. (1990). Immunological involvement in the efficacy of praziquantel. Exp Parasitol, 71(2), 245–248.

Brindley, P. J., Strand, M., Norden, A. P. & Sher, A. (1989). Role of host antibody in the chemotherapeutic action of praziquantel against Schistosoma mansoni: Identification of target antigens. Mol Biochem Parasitol, 34(2), 99–108.

Bueding, E., Liu, C. L. & Rogers, S. H. (1972). Inhibition by metrifonate and dichlorvos of cholinesterases in schistosomes. Br J Pharmacol, 46(3), 480–487.

Camacho, M. & Agnew, A. (1995a). Glucose uptake rates by Schistosoma mansoni, S. haematobium, and S. bovis adults using a flow in vitro culture system. J Parasitol, 81(4), 637–640.

Camacho, M. & Agnew, A. (1995b). Schistosoma: Rate of glucose import is altered by acetylcholine interaction with tegumental acetylcholine receptors and acetylcholinesterase. Exp Parasitol, 81(4), 584–591.

Camacho, M., Alsford, S. & Agnew, A. (1996). Molecular forms of tegumental and muscle acetylcholinesterases of Schistosoma. Parasitology, 112(Pt 2), 199–204.

Camacho, M., Alsford, S., Jones, A. & Agnew, A. (1995). Nicotinic acetylcholine receptors on the surface of the blood fluke Schistosoma. Mol Biochem Parasitol, 71(1), 127–134.

Campbell, W. & Garcia, W. (1986). Trematode infections of man. In W. Campbell & R. Rew (Eds.), Chemotherapy of Parasitic Diseases (pp. 392–393). New York: Plenum.

Caulfield, J. P., Chiang, C. P., Yacono, P. W., Smith, L. A. & Golan, D. E. (1991). Low density lipoproteins bound to Schistosoma mansoni do not alter the rapid lateral diffusion or shedding of lipids in the outer surface membrane. J Cell Sci, 99(Pt 1), 167–173.

de Mendonca, R. L., Beck, E., Rumjanek, F. D. & Goffeau, A. (1995). Cloning and characterization of a putative calcium-transporting ATPase gene from Schistosoma mansoni. Mol Biochem Parasitol, 72(1–2), 129–139.

Doehring, E., Poggensee, U. & Feldmeier, H. (1986). The effect of metrifonate in mixed Schistosoma haematobium and Schistosoma mansoni infections in humans. Am J Trop Med Hyg, 35(2), 323–329.

Doenhoff, M. J., Modha, J. & Lambertucci, J. R. (1988). Anti-schistosome chemotherapy enhanced by antibodies specific for a parasite esterase. Immunology, 65(4), 507–510.

el-Karaksy, H., Hassanein, B., Okasha, S., Behairy, B. & Gadallah, I. (1999). Human fascioliasis in Egyptian children: Successful treatment with triclabendazole. J Trop Pediatr, 45(3), 135–138.

Espinoza, B., Tarrab-Hazdai, R., Silman, I. & Arnon, R. (1988). Acetylcholinesterase in Schistosoma mansoni is anchored to the membrane via covalently attached phosphatidylinositol. Mol Biochem Parasitol, 29(2–3), 171–179.

Espinoza, B., Parizade, M., Ortega, E., Tarrab-Hazdai, R., Zilberg, D. & Arnon, R. (1995). Monoclonal antibodies against acetylcholinesterase of Schistosoma mansoni: Production and characterization. Hybridoma, 14(6), 577–586.

Fallon, P. G. & Doenhoff, M. J. (1994). Drug-resistant schistosomiasis: Resistance to praziquantel and oxamniquine induced in *Schistosoma mansoni* in mice is drug specific. *Am J Trop Med Hyg, 51*(1), 83–88.

Fallon, P. G., Fookes, R. E. & Wharton, G. A. (1996). Temporal differences in praziquantel- and oxamniquine-induced tegumental damage to adult *Schistosoma mansoni*: Implications for drug–antibody synergy. *Parasitology, 112*(Pt 1), 47–58.

Ferguson, M. A., Haldar, K. & Cross, G. A. (1985). *Trypanosoma brucei* variant surface glycoprotein has a sn-1,2-dimyristyl glycerol membrane anchor at its COOH terminus. *J Biol Chem, 260*(8), 4963–4968.

Fetterer, R. H., Pax, R. A. & Bennett, J. L. (1980). *Schistosoma mansoni*: Characterization of the electrical potential from the tegument of adult males. *Exp Parasitol, 49*(3), 353–365.

Goudot-Crozel, V., Caillol, D., Djabali, M. & Dessein, A. J. (1989). The major parasite surface antigen associated with human resistance to schistosomiasis is a 37-kD glyceraldehyde-3P-dehydrogenase. *J Exp Med, 170*(6), 2065–2080.

Graham, M. K., McGeown, J. G. & Fairweather, I. (1999). Ionic mechanisms underlying spontaneous muscle contractions in the liver fluke, *Fasciola hepatica*. *Am J Physiol, 277* (2 Pt 2), R374–383.

Hall, T. M., Joseph, G. T. & Strand, M. (1995). *Schistosoma mansoni*: Molecular cloning and sequencing of the 200-kDa chemotherapeutic target antigen. *Exp Parasitol, 80*(2), 242–249.

Harn, D. A., Gu, W., Oligino, L. D., Mitsuyama, M., Gebremichael, A. & Richter, D. (1992). A protective monoclonal antibody specifically recognizes and alters the catalytic activity of schistosome triose-phosphate isomerase. *J Immunol, 148*(2), 562–567.

Harris, A. R., Russell, R. J. & Charters, A. D. (1984). A review of schistosomiasis in immigrants in Western Australia, demonstrating the unusual longevity of *Schistosoma mansoni*. *Trans R Soc Trop Med Hyg, 78*(3), 385–388.

Hawn, T. R. & Strand, M. (1993). Detection and partial characterization of glycosylphosphatidylinositol-specific phospholipase activities from *Fasciola hepatica* and *Schistosoma mansoni*. *Mol Biochem Parasitol, 59*(1), 73–81.

Hockley, D. J. (1973). Ultrastructure of the tegument of *Schistosoma*. *Adv Parasitol, 11*, 233–305.

Hockley, D. J. & McLaren, D. J. (1973). *Schistosoma mansoni*: Changes in the outer membrane of the tegument during development from cercaria to adult worm. *Int J Parasitol, 3*(1), 13–25.

Hoffmann, K. F. & Strand, M. (1996). Molecular identification of a *Schistosoma mansoni* tegumental protein with similarity to cytoplasmic dynein light chains. *J Biol Chem, 271*(42), 26117–26123.

Hoffmann, K. F. & Strand, M. (1997). Molecular characterization of a 20.8-kDa *Schistosoma mansoni* antigen. Sequence similarity to tegumental associated antigens and dynein light chains. *J Biol Chem, 272*(23), 14509–14515.

Hrckova, G. & Velebny, S. (1997). Effect of praziquantel and liposome-incorporated praziquantel on peritoneal macrophage activation in mice infected with *Mesocestoides corti* tetrathyridia (Cestoda). *Parasitology, 114*(Pt 5), 475–482.

Hrckova, G., Velebny, S. & Corba, J. (1998). Effects of free and liposomized praziquantel on the surface morphology and motility of *Mesocestoides vogae tetrathyridia* (syn. *M. corti*; Cestoda: Cyclophyllidea) in vitro. *Parasitol Res, 84*(3), 230–238.

Ismail, M., Botros, S., Metwally, A., William, S., Farghally, A., Tao, L. F., Day, T. A. & Bennett, J. L. (1999). Resistance to praziquantel: Direct evidence from *Schistosoma mansoni* isolated from Egyptian villagers. *Am J Trop Med Hyg, 60*(6), 932–935.

Jiang, J., Skelly, P. J., Shoemaker, C. B. & Caulfield, J. P. (1996). *Schistosoma mansoni*: The glucose transport protein SGTP4 is present in tegumental multilamellar bodies, discoid bodies, and the surface lipid bilayers. *Exp Parasitol, 82*(2), 201–210.

King, C. H. & Mahmoud, A. A. (1989). Drugs five years later: Praziquantel. *Ann Intern Med, 110*(4), 290–296.

King, S. M. & Patel-King, R. S. (1995). The $M(r) = 8,000$ and 11,000 outer arm dynein light chains from *Chlamydomonas* flagella have cytoplasmic homologues. *J Biol Chem, 270*(19), 11445–11452.

Lanar, D. E., Pearce, E. J., James, S. L. & Sher, A. (1986). Identification of paramyosin as schistosome antigen recognized by intradermally vaccinated mice. *Science, 234*(4776), 593–596.

Linder, E. & Thors, C. (1992). *Schistosoma mansoni*: Praziquantel-induced tegumental lesion exposes actin of surface spines and allows binding of actin depolymerizing factor, gelsolin. *Parasitology, 105*(Pt 1), 71–79.

Liu, J. L., Fontaine, J., Capron, A. & Grzych, J. M. (1996). Ultrastructural localization of Sm28 GST protective antigen in *Schistosoma mansoni* adult worms. *Parasitology, 113* (Pt 4)(4), 377–391.

Mansour, J. M., McCrossan, M. V., Bickle, Q. D. & Mansour, T. E. (2000). *Schistosoma mansoni* phosphofructokinase: Immunolocalization in the tegument and immunogenicity. *Parasitology, 120*(Pt 5), 501–511.

Mansour, T. (1959). Studies on the carbohydrate metabolism of the liver fluke *Fasciola hepatica*. *Biochim Biophys Acta, 34*, 456–464.

Marshall, I. (1987). Experimental chemotherapy. In D. Rollinson & A. Simpson (Eds.), *The Biology of Schistosomes* (pp. 399–423). San Diego: Academic Press.

Matsumoto, Y., Perry, G., Levine, R. J., Blanton, R., Mahmoud, A. A. & Aikawa, M. (1988). Paramyosin and actin in schistosomal teguments. *Nature, 333*(6168), 76–78.

McLaren, D. J. & Hockley, D. J. (1977). Blood flukes have a double outer membrane. *Nature, 269*(5624), 147–149.

Mehlhorn, H., Becker, B., Andrews, P., Thomas, H. & Frenkel, J. K. (1981). In vivo and in vitro experiments on the effects of praziquantel on *Schistosoma mansoni*. A light and electron microscopic study. *Arzneimittelforschung, 31*(3a), 544–554.

Pajor, A. M., Hirayama, B. A., & Wright, E. M. (1992). Molecular biology approaches to comparative study of Na^+/glucose cotransport. *Am J Physiol, 263*(3Pt 2), R 489–95.

Pappas, P. (1988). The relative roles of the intestine and external surfaces in the nutrition of mongeneans, digeneans and nematodes. *Parasitology, 96*(Suppl), S105–121.

Parent, L., Supplisson, S., Loo, D. D. & Wright, E. M. (1992). Electrogenic properties of the cloned Na^+/glucose cotransporter: II. A transport model under nonrapid equilibrium conditions. *J Membr Biol, 125*(1), 63–79. [Published erratum appears in *J Membr Biol* 1992, *130*(2), 203.]

Redman, C., Robertson, A., Fallon, P., Modha, J., Kusel, M., Doenhoff, M. & Martin, R. (1996). Praziquantel: An urgent and exciting challenge. *Parasitol Today, 12*, 14–20.

Remold, H. G., Mednis, A., Hein, A. & Caulfield, J. P. (1988). Human monocyte-derived macrophages are lysed by schistosomula of *Schistosoma mansoni* and fail to kill the parasite after activation with interferon gamma. *Am J Pathol, 131*(1), 146–155.

Richman, D. P. & Arnason, B. G. (1979). Nicotinic acetylcholine receptor: Evidence for a functionally distinct receptor on human lymphocytes. *Proc Natl Acad Sci USA, 76*(9), 4632–4635.

Rogers, S. & Bueding, E. (1975). Anatomical localization of glucose uptake by *Schistosoma mansoni* adults. *Int J Parasitol, 3*, 369–371.

Sabah, A. A., Fletcher, C., Webbe, G. & Doenhoff, M. J. (1985). *Schistosoma mansoni*: Reduced efficacy of chemotherapy in infected T-cell-deprived mice. *Exp Parasitol, 60*(3), 348–354.

Sauma, S. Y. & Strand, M. (1990). Identification and characterization of glycosyl-phosphatidylinositol-linked *Schistosoma mansoni* adult worm immunogens. *Mol Biochem Parasitol, 38*(2), 199–209.

Shaw, M. K. & Erasmus, D. A. (1983). *Schistosoma mansoni*: Dose-related tegumental surface changes after in vivo treatment with praziquantel. *Z Parasitenkd, 69*(5), 643–653.

Skelly, P. J. & Shoemaker, C. B. (1996). Rapid appearance and asymmetric distribution of glucose transporter SGTP4 at the apical surface of intramammalian-stage *Schistosoma mansoni*. *Proc Natl Acad Sci USA, 93*(8), 3642–3646.

Skelly, P. J. & Shoemaker, C. B. (2001). The *Schistosoma mansoni* host- interactive tegument forms from vesicle eruptions of a cyton network. *Parasitology, 122*(Pt 1), 67–73.

Skelly, P. J., Kim, J. W., Cunningham, J. & Shoemaker, C. B. (1994). Cloning, characterization, and functional expression of cDNAs encoding glucose transporter proteins from the human parasite *Schistosoma mansoni*. *J Biol Chem, 269*(6), 4247–4253.

Stitt, A. W. & Fairweather, I. (1994). The effect of the sulphoxide metabolite of triclabendazole ('Fasinex') on the tegument of mature and immature stages of the liver fluke, *Fasciola hepatica*. *Parasitology, 108*(Pt 5), 555–567.

Stoitsova, S. R., Gorchilova, L. N. & Danek, J. (1992). Effects of three anthelmintics on the tegument of *Hymenolepis fraterna* (Cestoda). *Parasitology, 104*(Pt 1), 143–152.

Thompson, D. & Geary, T. (1995). The structure and function of helminth surfaces. In J. Marr & M. Muller (Eds.), *Biochemistry and Molecular Biology of Parasites* (pp. 203–232). San Diego: Academic Press.

Thompson, D. P., Pax, R. A. & Bennett, J. L. (1982). Microelectrode studies of the tegument and sub-tegumental compartments of male *Schistosoma mansoni*: An analysis of electrophysiological properties. *Parasitology, 85*(Pt 1), 163–178.

Threadgold, L. (1963). The tegument and associated structures of *Fasciola hepatica*. *Quant J Microsc, 104*, 505–512.

Threadgold, L. (1984). Parasitic platyhelminths. In J. Bereiter-Hahn, A. Matoltsy, & K. Richards (Eds.), *Biology of the Integument* (pp. 132–191). Berlin: Springer-Verlag.

Vasconcelos, E. G., Ferreira, S. T., Carvalho, T. M. U., Souza, W., Kettlun, A. M., Mancilla, M., Valenzuela, M. A. & Verjovski-Almeida, S. (1996). Partial purification and immunohistochemical localization of ATP diphosphohydrolase from *Schistosoma mansoni*. Immunological cross- reactivities with potato apyrase and *Toxoplasma gondii* nucleoside triphosphate hydrolase. *J Biol Chem, 271*(36), 22139–22145.

Webster, M., Fulford, A., Braun, G., Ouma, J., Kariuki, H., Havercroft, J., Gachuhi, K., Sturrock, R., Butterworth, A. & Dunne, D. (1996). Human immunoglobulin E responses to a recombinant 22.6-kilodalton antigen from *Schistosoma mansoni* adult worms are associated with low intensities of reinfection after treatment. *Infect Immun, 64*, 4042–4046.

WHO. (1992). Praziquantel shows unexpected failure in recent schistosomiasis outbreak. *TDR News, 41,* 2–3.

WHO. (1997). *Progress Report* p121. Geneva: WHO, Division of Control of Tropical Diseases.

Wolde Mussie, E., Vande Waa, J., Pax, R. A., Fetterer, R. & Bennett, J. L. (1982). *Schistosoma mansoni:* Calcium efflux and effects of calcium-free media on responses of the adult male musculature to praziquantel and other agents inducing contraction. *Exp Parasitol, 53*(2), 270–278.

Zhong, C., Skelly, P. J., Leaffer, D., Cohn, R. G., Caulfield, J. P. & Shoemaker, C. B. (1995). Immunolocalization of a *Schistosoma mansoni* facilitated diffusion glucose transporter to the basal, but not the apical, membranes of the surface syncytium. *Parasitology, 110*(Pt 4), 383–394.

Epilogue

One focus of this book is on parasitic targets that have been implicated in mechanisms of action of known chemotherapeutic effects and how this knowledge can be used in the rational search for new antiparasitic agents. A second focus is on ways to discover new targets among enzymes, in parasitic development, or thorough genomics. The twenty-first century marks a milestone in the development of many new technologies in medical research. It is timely that several international groups with financial support from WHO, the Wellcome Trust, and other funding agencies are sequencing and mapping the genomes of parasites that have had severe impacts on public health. Together with knowledge of the human genome and of *C. elegans*, the forthcoming information about the genomes of *P. falciparum, Trypanosoma cruzi* and *brucei, Leishmania major*, and *Brugia malayi* should greatly increase our ability to find selective targets and their inhibitors. Identification of parasitic genes, however, does not imply recognition of their functions. The number of genes that have identifiable functions are far fewer than those whose functions have not yet been identified. New technologies to simplify and hasten functional identification have not been sufficiently exploited for studies on parasites.

One important recent technique for studying gene function is post-transcriptional gene silencing. This has been used with great success in the nematode *C. elegans*. Synthetic double-stranded RNA (dsRNA), when introduced into the worm, is cleaved to yield small interfering RNAs (duplexes of 21–23 nucleotides), which target the corresponding mRNA and lead to its inactivation. For this reason the technique is also referred to as RNA interference (RNAi). Disruption of gene function by RNAi is very specific. Only genes that share sequence with the dsRNA are affected. Silencing the gene appears to last for the entire life cycle of the progeny of the treated organism. The molecular mechanisms of gene silencing are still being investigated. For a review see Fire (1999). RNAi technology has also been successfully used with fruit flies, trypanosomes, and plants. The technology for using RNAi in mammals is still being worked out and may have clinical implications in the future.

In addition to the use of RNAi in identifying the biological function of individual genes, the technology can also be used to identify genes involved in drug action or drug resistance, as well as in parasite development. For a review see Kuwabara & Coulson (2000). A distinctive use of the RNAi technique in *T. brucei* was reported in a study on the function of topoisomerase II and its relationship to protozoal development (Wang & Englund, 2001). Expression of a topoisomerase II dsRNA as a stem-loop caused a decrease in parasite growth, leading to parasite death. Silencing expression of this enzyme resulted in accumulation of minicircle replication intermediates and subsequent loss of the kinetoplast mitochondrial DNA network. These findings uncovered that a main function of topoisomerase II is to attach free minicircles to the mitochondrial DNA network. Because treatment of the parasite with dsRNA of topoisomerase II was eventually lethal to the parasite it is tempting to speculate that the procedure could be of clinical value as a treatment, provided it is used in a way that will not harm the host.

A new technology, DNA microarrays, allows simultaneous analysis of transcriptional responses in a large number of genes on a single array. The procedure is particularly useful for examining developmental regulation of specific RNAs. This technology has successfully been used to identify transcriptional profiles of human cells in response to *in vitro* infection with *Toxoplasma gondii* (Blader, Manger, & Boothroyd, 2001). Among the earliest genes expressed after infection were twenty-seven genes that encode for proteins associated with the immune response. These genes are also up-regulated by soluble parasite-derived factors produced before parasite invasion of the mammalian cells. Microarray analysis of fibroblast cells 24 hours after infection identified modulated host enzymes involved in glycolysis and mevalonate metabolism. These results lay the foundation for further studies on the nature of host–pathogen interaction.

Recent studies on filarial nematodes (*Brugia, Onchocerca, Wuchereria*) showed that these parasites harbor *Wolbachia* endosymbiotic bacteria. These bacteria appear to play an important role in the development, viability, and fertility of adult parasites. One feature of host response to filaria is that neutrophilic granulocytes attracted to the parasites initiate encapsulation of the adult parasites, thus forming the characteristic filarial nodules. This chemotactic response of the neutrophilic granulocytes toward the parasites is induced directly by products (which have not yet been identified) from the *Wolbachia* endobacteria (Brattig, Buttner, & Hoerauf, 2001). This contribution of the endosymbiotic bacteria can be stopped by treating the host with doxycycline or other broad-spectrum tetracyclines that kill the bacteria and stop the chemotactic response of the neutrophils. Initial usage of ivermectin against filarial infections (see Chapter 7) showed that the drug depletes skin microfilariae (produced by adult macrofilariae) for only a few months. However, the microfilariae reappear within a year. The reason is that ivermectin alone is not effective against adult parasites, which can live for as long as fifteen years and continue producing new microfilariae. Radical cure of filarial infections should include disappearance of the macrofilariae or the inflammatory filarial nodules that are necessary for the development of microfilariae to adults. Recently, treatment with a combination of doxycycline and ivermectin was found to deplete the *Wolbachia* endobacteria and the macrofilariae were shown to have produced no microfilariae when the patients were examined nineteen months after finishing the treatment. This shows cooperative effects between the two drugs to stop the production of microfilariae (Hoerauf *et al.*, 2001). The combined use of ivermectin and doxycycline also proved lethal to the macrofilariae of bovine onchocerciasis caused by *O. ochengi*. These results illustrate the point that antiparasitic targets could include symbiotic endobacteria as well as the parasite itself. More information is also needed regarding symbiotic bacteria of other helminth parasites and their role in parasite survival in the host. For a review see Hoerauf, Fleischer, & Walter, 2001.

Oxidative stress is considered to be significant for weakening or killing parasites. Inhibition of protozoal parasites' antioxidant mechanisms is a good strategy for selecting chemotherapeutic agents. Erythrocytic stages of malaria live in a milieu that is subject to oxidative stress. The parasites, therefore, have to depend on their own antioxidant defense mechanism, glutathione reductase, which is independent of the host's erythrocytic enzyme. Because the parasite enzyme is closely related to the erythrocyte enzyme, the parasite glutathione reductase has not been seen to be a good target for selecting antimalarial agents. Recently, kinetic characterization of glutathione reductase from *P. falciparum* showed that individual molecular species of the enzyme differ in their susceptibility to inhibitors. For example, the reductive form of the reaction is similar to the analogous reaction for glutathione reductase from the human host. However, the oxidative form of the reaction is biphasic, reflecting formation and breakdown of different molecular species of glutathione reductase that prevail in the cell. Enzyme inhibitors could, perhaps, be selected on the basis of glutathione reductase molecular forms that prevail in the parasite (Bohme et al., 2000).

Unlike malaria, trypanosomatids depend on trypanothione to handle the oxidants generated by their own metabolism (see Chapter 5). Compared to their host's glutathione, trypanothione has more reactive SH groups. There is also a significant difference between host and parasite in the downstream reactions of trypanothione to metabolize peroxides. Trypanosomatids do not have glutathione peroxidase, a host enzyme for the reduction of peroxides. Instead, the trypanothione is coupled to two enzyme systems, tryparedoxin and tryparedoxin peroxidase. Electrons flow from trypanothione to tryparedoxin and then to tryparedoxin peroxidase that catalyzes the reduction of hydroperoxides. For a review see Flohe, Hecht, & Steinert (1999). In addition to trypanothione reductase, attempts are being made to characterize tryparedoxin and tryparedoxin peroxidase as new possible targets. The crystal structure of tryparedoxin-I (in the oxidized state) from *Crithidia fasciculata* was reported by the Fairlamb and Hunter laboratories and compared with that of human thioredoxin (Alphey *et al.*, 1999). Among the reported differences between the crystal structures of the two enzymes is the content and orientation of helical components of secondary structures at the surface and around the active site. These indicate different shapes and charge distributions between the human and parasitic enzymes. Thus, tryparedoxin-I can be considered as a selective target for identifying inhibitors of hydroperoxide metabolism in trypanosomes. However, studies on the crystal structure of tryparedoxin peroxidase in the reduced state gave less indication of differences between parasite and mammalian enzymes at the active site, making it an unlikely target (Alphey *et al.*, 2000). This is a good example of the advantages of having the crystal structure of the target for identifying selective inhibitors.

Even though genomic sequences of most parasites are still incomplete the knowledge we have has significantly advanced our search for antiparasitic targets. The current efforts of highly accomplished scientists is enhancing the momentum in this distinctive field of research. Nonetheless, there are other important questions that should be tackled, such as determining the mechanisms by which parasites can alter their metabolic pathways when they encounter noxious chemical agents in their hosts. An example (given in Chapter 6) is the response of *Trichomonas vaginalis* to treatment with metronidazole. We know about some of the events involved, such as lowering gene expression of pyruvate:ferredoxin oxidoreductase and ferredoxin, two vital components of energy metabolism in these parasites. But, important and exciting as these discoveries are, they do not provide us with information about the signaling mechanisms underlying these responses in the parasites. Similar inquiries can also be made about changes in the energy metabolic pathways as the parasite shifts from the free-living stage to life in the mammalian host.

Purifying enzymes that catalyze rate-limiting reactions and studying their kinetics and regulatory properties are currently unfashionable. Although we are now able to identify every amino acid in an enzyme and in many cases its 3-D structure as well, we still need to know its kinetic characteristics, control mechanisms, and linkages to different signaling systems. In Chapter 5 an interesting finding regarding the mechanism of action of eflornithine, a prime antiprotozoal agent, was discussed. The antitrypanosomal effect is not because the drug has a lower K_i against the parasite ornithine decarboxylase than against the mammalian enzyme, but because the parasite enzyme has a much longer half-life than the mammalian enzyme.

A conclusion can thus be drawn that with the availability of parasite genomics and the contributions of traditional disciplines of biochemistry and parasite biology, the time has come that the burden of parasites will be exponentially reduced. We should always be prepared with alternative chemotherapeutic agents that can be used when older remedies lose their efficacy.

REFERENCES

Alphey, M. S., Leonard, G. A., Gourley, D. G., Tetaud, E., Fairlamb, A. H. & Hunter, W. N. (1999). The high resolution crystal structure of recombinant *Crithidia fasciculata* tryparedoxin-I. *J Biol Chem, 274*(36), 25613–25622.

Alphey, M. S., Bond, C. S., Tetaud, E., Fairlamb, A. H. & Hunter, W. N. (2000). The structure of reduced tryparedoxin peroxidase reveals a decamer and insight into reactivity of 2Cys-peroxiredoxins. *J Mol Biol, 300*(4), 903–916.

Blader, I. J., Manger, I. D. & Boothroyd, J. C. (2001). Microarray analysis reveals previously unknown changes in *Toxoplasma gondii*-infected human cells. *J Biol Chem, 276*(26), 24223–24231.

Bohme, C. C., Arscott, L. D., Becker, K., Schirmer, R. H. & Williams, C. H., Jr. (2000). Kinetic characterization of glutathione reductase from the malarial parasite *Plasmodium falciparum*. Comparison with the human enzyme. *J Biol Chem, 275*(48), 37317–37323.

Brattig, N. W., Buttner, D. W. & Hoerauf, A. (2001). Neutrophil accumulation around *Onchocerca* worms and chemotaxis of neutrophils are dependent on *Wolbachia* endobacteria. *Microbes Infect, 3*(6), 439–446.

Fire, A. (1999). RNA-triggered gene silencing. *Trends Genet, 15*(9), 358–363.

Flohe, L., Hecht, H. J. & Steinert, P. (1999). Glutathione and trypanothione in parasitic hydroperoxide metabolism. *Free Radical Biol Med, 27*(9–10), 966–984.

Hoerauf, A., Fleischer, B. & Walter, R. D. (2001). Of filariasis, mice and men. *Trends Parasitol, 17*(1), 4–5.

Hoerauf, A., Mand, S., Adjei, O., Fleischer, B. & Buttner, D. W. (2001). Depletion of wolbachia endobacteria in *Onchocerca volvulus* by doxycycline and microfilaridermia after ivermectin treatment. *Lancet, 357*(9266), 1415–1416.

Kuwabara, P. E. & Coulson, A. (2000). RNAi – Prospects for a general technique for determining gene function. *Parasitol Today, 16*(8), 347–349.

Wang, Z. & Englund, P. T. (2001). RNA interference of a trypanosome topoisomerase II causes progressive loss of mitochondrial DNA. *EMBO J, 20*(17), 4674–4683.

Index

Numbers in **Bold** indicate page where chemical formula is shown.